T0318970

Nature, Society, and Justice in the Anthropocene

Are money and technology the core illusions of our time? In this book, Alf Hornborg offers a fresh assessment of the inequalities and environmental degradation of the world. He shows how both mainstream and radical economists are limited by a particular worldview and, as a result, do not grasp that conventional money is at the root of many of the problems that are threatening societies, not to mention the biosphere. Hornborg demonstrates how market prices obscure asymmetric exchanges of resources – human labor, land, energy, materials – under a veil of fictive reciprocity. Such unequal exchange, he claims, underpins the phenomenon of technological development, which is, fundamentally, a redistribution of time and space – human labor and land – in world society. Hornborg deftly illustrates how money and technology have shaped our thinking and our social and ecological relations, with disturbing consequences. He also offers solutions for their redesign in ways that will promote justice and sustainability.

Alf Hornborg is an anthropologist and professor of human ecology at Lund University, Sweden. He is the author of *The Power of the Machine, Global Ecology and Unequal Exchange* and *Global Magic*. He has conducted field research in Canada, Peru, and Brazil.

NEW DIRECTIONS IN SUSTAINABILITY AND SOCIETY

Series Editors

JONI ADAMSON
Arizona State University
SHAUNA BURNSILVER
Arizona State University

Editorial Board

Published in conjunction with the School of Sustainability at Arizona State University and the Amerind Museum and Research Center in Dragoon, Arizona, New Directions in Sustainability and Society features a program of books that focus on designing a resilient and sustainable future through a rich understanding of past and present social and ecological dynamics. Collectively, they demonstrate that sustainability research requires engagement with a range of fields spanning the social and natural sciences, humanities, and applied sciences. Books in the series show that a successful transition to a sustainable future will depend on the ability to apply lessons from past societies and link local action to global processes.

For more information about the series, please visit http://newdirections.asu.edu/.

Titles in the Series

Nature, Society, and Justice in the Anthropocene

Unraveling the Money-Energy-Technology Complex

ALF HORNBORG

Lund University

CAMBRIDGE
UNIVERSITY PRESS

CAMBRIDGE
UNIVERSITY PRESS

University Printing House, Cambridge CB2 8BS, United Kingdom

One Liberty Plaza, 20th Floor, New York, NY 10006, USA

477 Williamstown Road, Port Melbourne, VIC 3207, Australia

314-321, 3rd Floor, Plot 3, Splendor Forum, Jasola District Centre, New Delhi - 110025, India

79 Anson Road, #06-04/06, Singapore 079906

Cambridge University Press is part of the University of Cambridge.

It furthers the University's mission by disseminating knowledge in the pursuit of education, learning and research at the highest international levels of excellence.

www.cambridge.org
Information on this title: www.cambridge.org/9781108454193
DOI: 10.1017/9781108554985

First published 2019
First paperback edition 2021

A catalogue record for this publication is available from the British Library

Library of Congress Cataloging in Publication data
NAMES: Hornborg, Alf, 1954– author.
TITLE: Nature, society, and justice in the anthropocene : unraveling the money-energy-technology complex / Alf Hornborg, Lund University, Sweden.
DESCRIPTION: Cambridge, United Kingdom ; New York, NY : Cambridge University Press, 2019. | Series: New directions in sustainability and society | Includes bibliographical references and index.
IDENTIFIERS: LCCN 2018058433 | ISBN 9781108429375 (hardback : alk. paper) | ISBN 9781108454193 (pbk. : alk. paper)
SUBJECTS: LCSH: Capitalism–Moral and ethical aspects. | Economic development–Moral and ethical aspects. | Environmentalism. | Social justice.
CLASSIFICATION: LCC HB501 .H6425 2019 | DDC 303.44–dc23
LC record available at https://lccn.loc.gov/2018058433

ISBN 978-1-108-42937-5 Hardback
ISBN 978-1-108-45419-3 Paperback

Cambridge University Press has no responsibility for the persistence or accuracy of URLs for external or third-party internet websites referred to in this publication, and does not guarantee that any content on such websites is, or will remain, accurate or appropriate.

"Have you ever wondered just what is the ultimate driving force behind the Anthropocene, the proposed new epoch in Earth history? It's our love of money. This book unravels the complex, profound force of money, not only to structure a global economy but also to destabilise the climate system and drive the Earth's biosphere towards its sixth great extinction event. Alf Hornborg's latest book is a must-read for anyone who wants to understand not only how a feature of everyday life that we all take for granted has the power to disrupt the functioning of an entire planet, but what we can do about it."

– Will Steffen,
Honorary Professor, Department of Geography and Geology, Copenhagen University; author of *Global Change and the Earth System: A Planet under Pressure* (2005)

"Alf Hornborg's *Nature, Society and Justice in the Anthropocene* throws down the gauntlet. The Anthropocene, he argues, exposes two of industrial capitalism's founding principles consolidated at the historical zenith of the British Empire: first, that economics is concerned with money and market prices, and need not consider nature; and second, that engineering harnesses natural forces with no need to consider the structure of world society. It is this distinctively modern form of thought that must be challenged, ontologically and politically. Curbing asymmetric global resource flows and climatic catastrophe will require less a tinkering with carbon emissions than a radical rethink and redesign of the artifacts – the money-energy-technology complex – of modernity. A bravura performance and endlessly provocative, *Nature, Society and Justice in the Anthropocene* is a book that commands our attention and a response."

– Michael Watts,
Class of 63 Emeritus Professor of Geography and Development Studies, University of California, Berkeley; author of *Silent Violence: Food, Famine and Peasantry in Northern Nigeria* (2005)

"Alf Hornborg's magnum opus is a bracing challenge to the two main streams of progressive critique of capitalism and the ecological crisis: the Marxist and the posthumanist. Both are going to have to engage seriously with Hornborg's argument, the essence of which is that power resides in the artifacts that rule out lives *and* that these artifacts – money and technology – are human constructions. Radical social change, including an adequate response to the unfolding catastrophe of the Anthropocene, requires that humans redesign those artifacts. It's a powerful, thought-provoking argument. If it's right, it points to a new direction out of the self-destructive impasse in which humanity finds itself."

– Clive Hamilton,
Professor of Public Ethics, Charles Sturt University; author of *Growth Fetish* (2003)

"Exploding illusions about the neutrality of technology, this innovative work centres on the role of money and technology in obscuring the fundamental inequalities in the flow of ecological resources that underpin capitalism and the global economic order. It should prove essential reading for all those seeking clarity on the injustice and environmental calamities of our times. Defending science and analytic rigour even as it gives an ear to their posthumanist critics and drawing on Marxism and Actor-Network Theory even as it quarrels with them, Hornborg invites us to think beyond conventional political understandings and their familiar categories. In exploring the confusions and evasions of current thinking on the nature–society divide, he opens up questions about value, progress, and the quest for wealth that are never asked by mainstream economists, and seldom even by their critics."

– Kate Soper,
Emerita Professor of Philosophy, London Metropolitan University;
author of *What Is Nature? Culture, Politics and the Non-Human* (1995)

"Cutting through the illusions that are blinding humanity to the reasons for our ecologically destructive trajectory while attacking the dithering of those who purport to be environmentalists but offer no real solutions, Alf Hornborg identifies the root cause of this trajectory and offers what is possibly the only practical way of altering it. He builds on, further defends and also goes beyond the arguments of his earlier work, most importantly *The Power of the Machine* and *Global Magic: Technologies of Appropriation from Rome to Wall Street,* in which he showed how the fetishism of commodities and of the technology it has generated have structured human relationships on a global scale to facilitate the maximum exploitation of labour and resources, particularly in the semi-peripheries and peripheries of the world economy. This is destroying not only these resources but the global ecosystem. In this work, he has shown how these fetishes are aspects of the development of general-purpose money with its tacit assumption that all values are commensurable. The solution is to recognize that money is an institution, and to institute different forms of money. A detailed solution is offered. Hornborg proposes a new form of money and shows how this could enable people to escape from the global system of exploitation and thereby avert global ecological destruction. He provides instruction on how to institute this money and how it would work. This book is a major advance in the struggle to save civilization from the collapse of the current regime of the global ecosystem."

– Arran Gare,
Associate Professor of Philosophy, Swinburne University of Technology,
Melbourne; author of *Postmodernism and the Environmental Crisis* (1995)
and *Philosophical Foundations of Ecological Civilization: A Manifesto for
the Future* (2016)

"For those concerned with blunting and reversing the burgeoning challenges to biospheric sustainability inherent in the dynamics of the Anthropocene, Alf Hornborg's deep dive into social theory is a must-read. Through a critique not just of the fields of economics and engineering and their contributions to our dilemma but also of Actor Network Theory, posthumanism, environmental humanities, and ecological Marxism, Hornborg lays the ontological groundwork for his own proposal that focuses on redesigning the human artifact of money as a way out of the crisis of the Anthropocene. Along the way, readers will be treated to Hornborg's engagement with and critique of other scholars debating the Anthropocene, and to his insights on the continuing relevance of Karl Polanyi, the global nature of technology, the flaws in proposals to substitute biofuels for fossil fuels, animism, and the agency (or not) of objects in the 'Human Era.'"

– Robert B. Marks,
Professor of History, Whittier College; author of *The Origins of the Modern World: A Global and Ecological Narrative from the Fifteenth to the Twenty-First Century* (2007)

"You may or may not agree with his proposal for 'redesigning money' to achieve community sufficiency, but Alf Hornborg's thesis in *Nature, Society, and Justice in the Anthropocene* is stimulating and provocative. He argues that while global socio-ecological relations today are organised by 'money' and 'technology', each is simply a fetishised 'cultural artifact'. Economics operates in a sealed-off idealised world with no capacity to engage the thermodynamic materiality of life on Earth. Likewise, people treat technology as 'a given' rather than a creation of a certain socio-historical formation. The book is clearly critical of neoliberal capital, but the author's anthropological experience distances him from some eurocentric assumptions in Marxism. Equally, he has a mixed response to analyses framed by Latour's Actor Network Theory. Hornborg's text speaks to the concerns of environmental economists, students of political economy, activists, and policy advisors, and is a welcome contribution to eco-socialist debates."

– Ariel Salleh,
University of Sydney; author of *Ecofeminism as Politics: Nature, Marx and the Postmodern* (1997)

"This book is ecological economics at its best. It explains the links between energy, money and growth and it makes a passionate call for redesigning money."

– Giorgos Kallis,
Professor of Environmental Science and Technology, Autonomous University of Barcelona; author of *Degrowth* (2018)

"A sweeping and deeply original analysis of how universal money has interpenetrated and mediated the modern coevolution of social and natural worlds. Alf Hornborg critiques the failure of local and regional analyses of technology and trade and takes on global analyses as well. Addressing unconstrained universal money is key to escaping the rush toward greater inequality and human suffering, climate change, and the sixth extinction."

– Richard B. Norgaard,
Professor Emeritus of Energy and Resources, University of California,
Berkeley; author of *Development Betrayed: The End of Progress and a Coevolutionary Revisioning of the Future* (1994)

"With money, capitalism, Marxism, political ecology, environmental degradation, and justice as anchoring themes, Hornborg ranges widely across many aspects of current anthropology (including its fringes). The book is important reading for scholars of these topics and for all concerned about the future of humanity and the earth. Hornborg proposes that problems of the environment and justice call for a redesign of money for local use only within each nation, existing alongside regular currencies."

– Joseph Tainter,
Professor, Department of Environment and Society, Utah State University;
author of *The Collapse of Complex Societies* (1988)

In memory of my brother Sten Eirik (1952–2015)

Do not go gentle into that good night.
Rage, rage against the dying of the light.
Dylan Thomas

Contents

Figures

Acknowledgments

For permission to include revised versions of texts previously published elsewhere, I thank the following journals and publishers:

1 "Fetishistic Causation: The 2017 Stirling Lecture," *HAU: Journal of Ethnographic Theory* 7(3): 89–103 (2017).

4 "The Root of All Evil," in *Market versus Society* (Palgrave Macmillan, 2018), edited by Manos Spyridakis, reproduced with permission of Palgrave Macmillan.

7 "The Magic of Money and the Illusion of Biofuels," *European Physical Journal Plus* 132(82): NA (2017).

8 "Energy, 'Value' and Fetishism in the Anthropocene," *Capitalism Nature Socialism* 27(4): 61–76 (2016).

9 "The Money-Energy-Technology Complex and Ecological Marxism," *Capitalism Nature Socialism*, https://doi.org/10.1080/10455752.2018.1440614 (2018).

10 "Artifacts Have Consequences, Not Agency," *European Journal of Social Theory* 20(1): 95–110 (2017).

11 "Dithering While the Planet Burns," *Reviews in Anthropology* 46 (2–3): 61–77 (2017).

12 "Convictions, Beliefs, and the Suspension of Disbelief," *HAU: Journal of Ethnographic Theory* 7(1): 553–558 (2017); "Relationism as Revelation or Prescription?," *Interdisciplinary Science Reviews* 43(3–4): 253–263 (2018).

13 "How to Turn an Ocean Liner," *Journal of Political Ecology* 24 (1): 623–632 (2017).

I thank Sage Publications for permission to reproduce Figure 3a, published as Map 3 (GDP Density) in J. L. Gallup, J. D. Sachs, and A. D. Mellinger, "Geography and Economic Development," *International Regional Science Review* 22(2): 179–232 (1999). Figure 3b is reproduced courtesy C. Mayhew and R. Simmon, NASA.

At Cambridge University Press I want to thank Joni Adamson, Katherine Barbaro, Christine Dunn, Mark Fox, Edgar Mendez, Beatrice Rehl, Mathew Rohit, and Norman Yoffee for their efforts.

The arguments and ideas in this book have grown out of many years of engaging conversations with a far-flung network of colleagues in anthropology, geography, economics, sociology, history, archaeology, and several other fields. I owe these colleagues a debt that goes back many more years than it has taken me to write this book. For once I will embark on a fairly long list, although I am aware of the risk of forgetting to include people who should be on it.

For inspiration and stimulating discussions on different occasions, I am indebted to Zygmunt Bauman†, Christophe Bonneuil, Stephen Bunker†, Brett Clark, Philippe Descola, Paul Durrenberger, Thomas Hylland Eriksen, John Bellamy Foster, Andre Gunder Frank†, Maurice Godelier, David Graeber, Stephen Gudeman, Torsten Hägerstrand†, Clive Hamilton, Keith Hart, Dougald Hine, Matt Huber, Tim Ingold, Giorgos Kallis, Keir Martin, Joan Martinez-Alier, Timothy Mitchell, Patrick Neveling, Stephen Nugent†, Bryan Pfaffenberger, Skip Rappaport†, Ariel Salleh, Frank Salomon, Sandy Smith-Nonini, Will Steffen, Bron Szerszynski, Joe Tainter, Eduardo Viveiros de Castro, Rick Wilk, and Richard York.

I have also benefitted from much appreciated collaboration and conversations with Miguel Alexiades, Eva Alfredsson, Jan Otto Andersson, Jonas Anshelm, Marco Armiero, Stefania Barca, Karin Bradley, Wim Carton, Gustav Cederlöf, Carole Crumley, Christian Dorninger, Jorge Gómez-Paredes, Kenneth Hermele, Joe Heyman, Jonathan Hill, Christian Isendahl, Anne Jerneck, Andrew Jorgenson, Kilian Jörg, Kristian Kristiansen, Stefano Longo, Tom Love, Mikael Lövgren, Mikael Malmaeus, John McNeill, Felipe Milanez, Janken Myrdal, Gunilla Almered Olsson, Lennart Olsson, Martin Oulu, Timothée Parrique, Susan Paulson, Daniela Peluso, Johannes Persson, Vasna Ramasar, Andreas Roos, Stephanie Rupp, Fernando Santos-Granero, Jonathan Seaquist, Michael Sheridan, Simron Jit Singh, Alevgul Sorman, Sarah Strauss, Paul Trawick, Rikard Hjorth Warlenius, and Norman Yoffee.

For highly valued friendship, encouragement, and dialogue over many years, I want to thank Kaj Århem, Pelle Brandström, Anders Burman, Eric

Clark, Gudrun Dahl, Kajsa Ekholm Friedman, Jonathan Friedman, Eva Friman, Pernille Gooch, Thomas Håkansson, Johan Hedrén, Cindy Isenhour, Beppe Karlsson, Mikael Kurkiala, Karl-Johan Lundquist, Andreas Malm, Thomas Malm, and Mats Widgren.

I thank all these friends and colleagues for sharing with me their concerns and struggles. And, as always, I owe more than I can express to my wife Anne-Christine, my children, and my grandchildren.

Introduction

This book is ultimately an argument for *degrowth*. However, it differs from other critiques of economic growth and capitalism in not approaching the imperative of growth as an *idea* or *policy* that we might argue about, as if better ideas and policies could be implemented on the basis of empirical evidence, reason, morality, democratic decision making, or political activism. All these avenues have been tried in innumerable debates and political experiments throughout the world – but all have finally proven futile. Everywhere politics have succumbed to the logic of the market. Democratic political systems have been unable to curb the logic of globalized capitalism. Democracy sees to it that any politician sincere enough to seriously advocate degrowth will not have a future in politics. The ecologically suicidal trajectory of global society is not the result of misguided policies, corrupt politics, or human nature, but of the imperatives inscribed in the artifact of general purpose money – the idea that everything can be converted into anything else.[1] Without redesigning

---

[1] In economic anthropology, general purpose money is distinguished from special-purpose money, which can only be used for transactions within a restricted sphere of exchange (cf. Bohannan 1955). Rather than refer to money as a general and universally homogeneous phenomenon, we need to "distinguish between those characteristics of money that are unique to capitalism and the various money forms (like cowrie shells or wampum beads) that pre-existed it" (Harvey 2018: 52). Although various forms of money have emerged over the millennia (cf. Weatherford 1997; Graeber 2011; Le Goff 2012), the establishment of modern, general purpose money accompanied the expansion of market trade in the eighteenth century. I may henceforth sometimes refer to general purpose money simply as money.

money, all our insights and aspirations for a better world will come to nothing. This is tantamount to proposing that we need to transform our very conception of politics by recognizing the power of artifacts to organize social life and, concomitantly, that to change society we need to redesign our artifacts.

The argument for degrowth is necessarily also an argument against capitalism. The case against capitalism has been made by countless writers, activists, and warriors over the past two centuries. Entire libraries can be filled with their analytical deliberations and manifestos and the accounts of their struggles. The case against growth did not begin mounting until the early 1970s, after two and a half postwar decades of unprecedented economic expansion primarily in North America, Europe, and Japan. Whereas anticapitalists continued to focus on the widening inequalities and injustices generated by the modern economy, the critique of growth added to this focus on unequal distribution an emphasis on how it tends to degrade the global environment. These two objections to what mainstream economists and most people in general think of as progress and development have long occupied center stage in political debates. They have organized the political arena in terms of opposed ideologies regarding key topics such as the benefits of unregulated market trade versus government intervention. Critics of capitalism and growth have tended to assume that this is a battle of *ideas* – that proponents of capitalism and growth suffer from misconceptions or ulterior motives that they might be persuaded to abandon. But my argument in this book is that advocacy of economic growth is *not* simply a conspiracy or a misconceived idea. The economists and the capitalists are not mistaken about how to most efficiently manage money – they know exceedingly well how money operates, what kinds of incentives it tends to generate among people, and how people will tend to pursue those incentives. General purpose money, however, *is* indeed a debatable idea – and growth is its implicit, built-in imperative because it inevitably generates incentives to conduct exchanges that augment the sum of money available at the outset and because growing sums of money represent growing claims on other people's labor and resources. The ultimate adversary of critics of the modern economy is not a group of misguided or evil people – capitalists, economists, or neoliberal politicians – but an *artifact* with an inherent inertia. While most political debate focuses on how money should be handled – to ensure not only growth and security but ostensibly also sustainability and resilience – I will argue that it is more essential to debate how money should be *designed*.

The movement now united under the banner of "degrowth" represents the recognition that the anticapitalist programs of socialism have not decreased global inequalities or environmental degradation. It signifies a widespread transformation of leftist sentiments particularly following the collapse of the Soviet Union. Rather than celebrate the triumph of capitalism and free market trade, it retains an anticapitalist critique but pursues even more radical ways of challenging business as usual than conventional socialist programs. The advocacy of degrowth in recent decades reflects the fact that the alternatives to capitalism offered by the political history of the twentieth century proved no better than the system they professed to replace. It is no coincidence that the calls for degrowth have attracted growing support during the decades that have seen the dissolution of the Soviet Union, the economic ascendancy of China, and the destinies of other socialist nations such as Venezuela and North Korea.

Looking back at the framing of public discourse, the early 1970s in Europe and North America were a time of traumatic disillusionment. The optimism of the postwar decades had been replaced by a gloomy understanding of global capitalism and a dystopian anticipation of crisis. The economic expansion of the 1950s and 1960s had fostered general optimism even as the environmental movement began questioning the foundations of industrial society. The message of *Silent Spring* (Carson 1962) had been a wake-up call, ominous but still possible to handle by a credible, modernist political establishment. The Vietnam War raised critical questions about the role of economically developed nations in an increasingly unequal world order, but even these doubts had seemed surmountable by a new and emancipated generation championing civil rights and announcing the arrival of the Age of Aquarius. However, a few years after the vociferous student revolt of 1968, the optimism of the various social movements that had been so diagnostic of modernity decisively gave way to postmodern hesitations about the economy, the future, and even our construction of reality. In 1971, excessive military expenditures in Vietnam forced President Nixon to abandon the Bretton Woods gold standard, leaving the dollar – and so much else – without a solid referent. The same year saw not only the birth of NASDAQ and electronic money but also the publication of several foundational critiques of the industrial economy, which up to then had been growing incessantly for a quarter of a century. This was the year Nicholas Georgescu-Roegen published *The Entropy Law and the Economic Process* (1971) and Howard Odum *Environment, Power and Society* (1971). The following year saw the publication of the Club of Rome report *The Limits to Growth* (Meadows et al. 1972), Arghiri Emmanuel's (1972) classic analysis of the imperialism of international trade, and the United Nations' Stockholm Conference on Environment and Development. The same year André Gorz coined the word *degrowth*, Eric Wolf (1972) launched the concept of "political ecology," and Gregory Bateson (1972) published *Steps to an Ecology of Mind*. Then, in 1973, came the first global oil crisis and the publication of E. F. Schumacher's *Small Is Beautiful: A Study of Economics as if People Mattered* (1973) and Ivan Illich's *Tools for Conviviality* (1973). In the years 1971 to 1973, in other words, a very widespread reappraisal of economic growth, technological development, and the reliance on fossil fuel energy shook Euro-American society.[2] While economic growth and the concomitant expansion of societal metabolism today continue to be promoted as the supreme purpose of human societies, we are still struggling to make sense of that traumatic loss of modernist self-confidence almost 50 years ago. We continue to be torn between the diametrically opposite perspectives of, for instance, Serge Latouche (2009) and Steven Pinker (2018).

Current deliberations on sustainability suggest a frustrating impasse. In this book I will argue that the fundamental categories of modern thought, aligned as they are with the features that we have attributed to the artifacts that organize our economy, are a common denominator of a myriad social and ecological problems experienced by people worldwide. But to fathom how deeply misguided we are by these categories, we need to keep a critical distance to the profusion of concepts that are now being generated to instill faith in "green" growth, sustainable development, ecological modernization, dematerialization, fair trade, circular economies, transition towns, and a host of other notions projecting the illusion that we are, after all, on the right track.

To the extent that we are *not* on the "right track," we need to establish what is the ultimate nature of our problem. Many would be content with responding "growth," suggesting an idea or policy that must be put into question. Even more would say "capitalism," as a shorthand for an abstract *system* that has been scrutinized and criticized ever since the pioneering analysis of Karl Marx. Others might focus on "globalization," the "market," "neoclassical economics," or even "modernity." While

---

[2] In his posthumously published book *Good Work*, Schumacher (1979) specifies "the date when a hole appeared in the skin of the balloon" as October 6, 1973. This was the date when it became evident to the world that economic growth was contingent on oil prices.

I have the utmost respect for all these voluminous discourses, I believe that the phenomena with which they are concerned are all the results of a more fundamental problem, which tends to remain the elephant in the room, namely, the idea and artifact of *general purpose money*: the assumption that *anything you have can be exchanged for anything else.*[3] This is so "natural" to us that we don't even see it. It is like water to fish. But no other species could have come up with such a strange notion. Nor is any other species projected to make the planet uninhabitable. I would not hesitate to claim that this is not a coincidence.

To grasp how money indeed is "the root of all evil," as St. Paul recognized more than two thousand years ago, we need to accept the claim of so-called Actor-Network theorists that artifacts intervene in social life. They don't have *agency* – that would be to fetishize them – but they have very significant consequences. Artifacts are invented by humans, but then we let them take control over us. We delegate the destiny of world society and the biosphere to these *things* and *ideas* of ours. This is what Karl Marx aptly classified as *fetishism*. Our tools become our masters. They compel us to behave in certain ways. The aggregate logic of these compulsions – these imperatives that are inscribed in the inertia of money – *is* capitalism. We all know that it is fundamentally about greed: everyone wants to pay as little as they can for what they buy and charge as much as they can for what they sell.[4] This logic leads inexorably to globalization, encouraging lower wages and more lax environmental legislation in the periphery of the world-system. The logic of the world market sees to it that low wages and lax environmental legislation are comparative advantages favoring imports from some countries. The ascent of China has been founded on the recognition that low wages are good for business. While the increasingly globalized logic of the market continues to unfold, economists are compelled to use a neoliberal discourse to justify these processes as "efficient" and "rational."

---

[3] This claim is often dismissed as reductionism. Thus, for example, the editor of an online publication advocating a radical transition of global society rejected it by arguing that his premise was instead "the emergence of a globalized social-ecological system conditioned by multiple co-causal, co-evolutionary developments and forces not reducible to single primary drivers." My rejoinder would be to ask if any of these "developments and forces" would be thinkable without general purpose money. Like the blind men trying to visualize the elephant, each specialized discourse will tend to defend its unique perspective.

[4] Greed has become "second nature" to modern existence. We have become accustomed to a range of corporate strategies to sell us things we don't need or that last ever shorter spans of time, prompting ever quicker rates of replacement.

Crucially, their preoccupation with monetary values obscures the asymmetric flows of biophysical resources (including embodied labor) that are orchestrated precisely by the market prices that they study. The real-life processes of the market and the ostensibly neutral observations of mainstream economists thus mutually reinforce each other. The logic of general purpose money generates not only unevenly distributed growth, globalization, increasing inequalities, and environmental degradation but also the social condition of modernity. This has long been obvious to philosophers and sociologists of money (e.g., Simmel 1990 [1907]). At an existential level, money conditions us to abstraction, interchangeability, and disembeddedness, which tends to alienate us not only from fellow humans but also from our natural environment. The history of ideas in sociology and environmental history is replete with observations to this effect.

The inherent potential of artifacts to mold our relations is evident to anyone who has ever played a game of *Monopoly*. For most of the players the game will inevitably end in disaster. The world economy can be metaphorically viewed as a gigantic board game. The winners don't win because they are evil. If that was the problem, it would make things much easier. It would make the politics of challenging inequalities and unsustainability into a simple matter of restraining the agency of evil people. But the rules of the game survive the substitution of human players. Unlike natural laws such as the principles of thermodynamics, those rules have been written by humans. They are social constructions. Even the pieces in the game – the checkers, so to speak – have been designed by humans. Yet it is symptomatic of fetishism that we find it hard to even imagine that those pieces can be redesigned, and the rules rewritten. This is clearly a psychological predicament: we need to be able to see alternatives to general purpose money before we are prepared to accept that the problem is how money is designed.

Our lives have been governed by general purpose money for only a few centuries, but humans have been here for hundreds of thousands of years, without destroying the biosphere. What can we learn from our history? What is it about general purpose money that, within a few centuries, brings us to the brink of disaster? I think a clue to the answer can be found in how our dictionaries define the word *liquidate*. Among alternative meanings, my Webster's dictionary provides "to convert into cash" and "to destroy." This is significant. In making all values interchangeable, general purpose money dissolves the kinds of distinctions on which all living systems depend: between the short term and the long term, the small scale and the large scale, the trivial and the essential. It makes it possible to trade Amazonian rainforests for Coca-Cola and the lives of African children for dividends on Wall Street. If we really want the kind of economic and ecological restraints evoked by visions of degrowth or a postcapitalist society we must redesign money. It is my conviction that there is no other way, short of disaster. That is why, in this book, I not only spell out the implications of such an approach for a wide range of topics – economic history, climate change, our concept of technology, the role of energy in human societies, our understandings of value and exploitation, and ideas on whether nonhuman objects have agency – but also offer a concrete proposal on how money and the economy could be redesigned.

In a nutshell, my argument begins with the phenomenon of general purpose money – an artifact of the uniquely human capacity for abstract symbolic representation – which generates the globalized market that has become the study object of mainstream economics. General purpose money has been conducive to the commodification of human time and natural space while obscuring material asymmetries in exchange and thus promoting time-space appropriation and other forms of unequal exchange. Capitalism is the aggregate logic of general purpose money. Major critics of this logic include Marx and Polanyi, but both failed to see that modern technology is contingent on unequal exchange. Marx believed that technology could be excised from capitalism, rather than intrinsically being a zero-sum game, in which some people save time and space at the expense of time and space lost for others. The main purpose of this book is to show how modern conceptions of money, energy, and technology serve as an ideology that obscures material processes of appropriation and exploitation. This ideology, moreover, tends to pervade the conceptual frameworks of both the guardians and the critics of business as usual.

A fundamental and insidious dilemma is that even the most radical critics of capitalism tend to frame its contradictions in terms of concepts that ultimately derive from its own assumptions. There is a widespread consensus among Marxists, widely defined, that the global economy for centuries has been degrading the environment (e.g., Bunker 1985; Foster et al. 2010; Moore 2015) – and some have suggested that the process can be understood in terms of "ecological unequal exchange" (Foster and Holleman 2014; Holleman 2018) – but these empirically rich interpretations of our global ecological predicament all tend to be couched in the notion that the root problem is an asymmetric transfer of underpaid "values" from periphery to core. Beginning with my publications on this topic in the 1990s, my argument is instead that ecologically unequal exchange is not about asymmetric transfers of values, but of resources. The transfers of resources are *orchestrated* by attributions of economic value, but our conceptualization of them must detach itself from theories of value. This is not a trivial quibble, because labor or energy theories of value are very unlikely ever to be taken seriously by the mainstream economists who continue to shape the dominant discourse on the relation between ecology and economy. To posit the existence of values that systematically contradict the operation of the market will not persuade mainstream economists. Not that I think they are likely to ever recognize unequal exchange in this sense either, but, unlike contested notions of purportedly "real" values, the acknowledgment that the market is degrading the biosphere is ultimately incontrovertible (cf. Georgescu-Roegen 1971). It is thus more analytically robust to argue that the market is "killing the planet" (Koumoundouros 2018) than to debate whether its assessments of "value" are justified. Even the economists will sooner or later have to take that argument seriously. Rather than champion contested understandings of value – which, as I show particularly in Chapter 9, ultimately derive from our immersion in the conceptual constraints of general purpose money – our task must be to show how market attributions of economic value inexorably lead to asymmetric transfers of resources. These asymmetric transfers of resources are in turn inextricably linked to increasing global inequalities and ecological degradation. They are also prerequisite to "development" – understood as the accumulation of technological infrastructure – which, paradoxically, is widely understood as something that might alleviate economic inequalities and environmental harm.

I was originally trained in anthropology, but most of the literature listed at the end of this book will be unfamiliar to anthropologists. For decades I have had the privilege of forging a transdisciplinary research field on the challenges of global justice and sustainability – we have called it *Human Ecology* – but I believe that transdisciplinary thinking is ultimately a personal endeavor. It can at times be an exhilarating pursuit, even if it is often agonizing and fragmenting. The fields drawn upon in this book include anthropology, history, economics, the philosophy of technology, energy transitions, environmental justice, industrial ecology, Marxist theory, Actor-Network Theory, and much more. I cannot, of course, do justice to any of them. But instead of apologizing for my attempts to bring together these disparate discourses, I want to emphasize that our only chance of grasping the predicament that is now being called the Anthropocene is to transcend our particular frames of reference and exert ourselves to connect divergent perspectives into a fundamentally revised worldview.

It is quite possible and not uncommon to spend a lifetime unraveling the conceptual framework of a single human mind – such as a writer, a philosopher, or one of the apostles – but we may sometimes need to remind ourselves that the meticulous exegesis of the thought of a person such as Karl Marx or St. Paul risks having but a tenuous relation to the real world in which we are immersed. When the anthropologist Claude Lévi-Strauss explored the structure of indigenous Amerindian thought systems he was convinced that his own mind was not only retracing those of Native Americans, but that such mental worlds in some way reflected nature – the biophysical world that had generated the human brain. He may have been right about the congruity or at least compatibility between Amerindian worldviews and natural conditions. But the impasse of the Anthropocene forces us to concede that there are aspects of our own, modern thought systems that very poorly reflect the biophysical world in which we are immersed. This book attempts to address this discrepancy between modern thought and social organization, on the one hand, and the natural, material conditions of our existence, on the other. Its main thesis is that fundamental aspects of the modern worldview have been shaped by our historical experience of two kinds of human artifacts – products of the human mind that have themselves reconfigured that mind to the point that its incentives and aspirations starkly contradict the biophysical conditions of human existence.

The first artifact that so pervasively leads our thoughts astray is money – or, more specifically, the so-called general purpose money that has increasingly come to dominate our lives over the past three centuries. Its basic idea is that most things are interchangeable on the market. This idea has transformed our ways of thinking, our social relations, and our relations to the natural world of which we are a part. It has made possible the kinds of modern technologies that require continuous inputs of fuels and other resources that can be purchased on the market, provided that the prices are right. To apply concepts familiar to the economists, the generalized fungibility of commodities on the global market has paved the way for technological development as made feasible by arbitrage.

This brings us to the second kind of artifact that has led our thoughts astray. The tangible materiality of technology misleads us into assuming that its existence is simply a matter of discovering how to assemble components of nature, as if access to those components were not a matter of *social* relations and rates of exchange. As I hope to show in this book, we need to fundamentally revise our understanding of the conditions of technological progress. But to grasp the inherent, distributive dimension of modern technology, we need to understand the role of money in obscuring what I have called *ecologically unequal exchange*.

These are convictions that have haunted me for a long time. If the reader will feel that I am repeating myself in the chapters of this book, it is because I have been intent on illuminating these basic conditions from a wide range of perspectives: theories of magic and fetishism, semiotics, world-system analysis, the history of economic thought, philosophy of technology, theories of energy transitions, thermodynamics, Marxist theory, Actor-Network Theory, and more. The transdisciplinary character of my argument is thus not just a methodological ambition but also a consequence of having been drawn into a very diverse set of conversations. To curb asymmetric global resource flows, and to avoid the most disastrous scenarios of the Anthropocene, our only chance is to critically rethink and redesign the artifacts – money and technology – that currently rule our thoughts and lives. It is ultimately money – and the technologies it makes possible – that is producing obscene social injustices and destroying the biosphere. I do not think that it is productive to blame a certain category of people such as capitalists or economists – or even an abstract system called *capitalism* – for these destructive processes. Indignation will not suffice to curb the exploitative and disastrous trajectories of general purpose money and capital accumulation, as illustrated historically by the destinies of movements such as Luddism or communism. The evils of the Anthropocene do not emerge from the character of any specific group of humans, but from our vehicles of interaction. For humans to assume responsibility for the future of society and the biosphere, we must be prepared to rethink how the checkers are designed in the game through which we engage each other.

The primary aim of this book is to challenge the understandings of money and technology that dominate mainstream thinking in economics and engineering, but a no less important point is its observation that even the most influential critiques of the current world order tend to be either constrained by such mainstream understandings or deluded by the complete rejection of modern analytical thought. Different chapters thus engage the shortcomings of Marxist theory and posthumanism, respectively. The most troubling impasse of the Anthropocene is the incapacity of its most radical critics to think beyond diagnoses such as "capitalism" or "dualism." As long as the ways in which nature and society are interfused in our artifacts are misunderstood – either by keeping them ontologically insulated from each other or by dissolving the analytical distinction between them – neither approach will grasp the dilemma of the Anthropocene. The reliance on fossil energy is no less massive in purportedly socialist countries such as China or Cuba than in mainstream capitalist countries, while the posthumanist abandonment of analytical distinctions between nature and society only obscures our global predicament. I hope to show that much Marxist theory remains locked within the walls of the same conceptual prison house that constrains the human targets of its critique, but that the solution cannot be to give up all hope of analytically unraveling how that prison house is constructed. The Marxist concept of "productive forces" is as entrenched in machine fetishism – as deludingly sequestered from global political economy – as the mainstream notion of technology. This is ironically illustrated by the observation that Cuba's post-Soviet ambition to shift to solar power has been constrained by a lack of *capital*. However, to understand why Cuba has not become the solar utopia envisaged by Schwartzman (1996),[5] the least enlightening approach we could choose would be to adopt Bennett's (2010) injunction to attribute agency to the energy technologies.

If we are concerned about global justice and sustainability, we must be prepared to deconstruct some central, modern ideas about progress, monetary value, and technology that tend to constrain orthodox and heterodox worldviews alike. But it would be counterproductive to let our concessions to deconstruction and relativism lead to a complete jettisoning of modern science and rationality, as appears to be the conclusion of some prominent proponents of the so-called ontological turn in anthropology, who tend to champion nonmodern worldviews and turn their backs on the Enlightenment. Very briefly, the common denominator of arguments said to represent an ontological turn is that people in different cultural contexts literally live in different worlds, and that their ways of understanding their realities are all equally valid in those contexts. For example, if an abiotic object is perceived by some nonmodern, indigenous group as animate and purposeful, it is held that the only reason why external observers such as European anthropologists might be skeptical to such claims is their own ethnocentrism. A corollary of this view is that Enlightenment rationality is a dubious imposition affiliated

---

[5] In 2015, Cuban electricity production remained dependent on fossil energy to about 95 percent.

with colonial power structures. My objections to some proponents of this approach are presented in Chapters 10, 11, and 12. Although I very much concur with the imperative of confronting neocolonialism and deconstructing the neoliberal understanding of progress and development through which it is reproduced, I must reject so-called posthuman propositions that artifacts, rivers, or mountains – or indeed a warming planet – have agency. I am unable to accept that such nonmodern propositions could provide valid arguments with which to politically confront the ecological degradation generated by modern economics and technology. In Chapter 12, I thus consider the problematic relation between posthumanism and political ecology.

To avoid the biophysical disaster that is being generated by business as usual, we shall certainly have to undergo an ontological cataclysm.[6] But this is where it is crucial to find a balance between radical deconstruction and rigorous analytical thought. I argue for a *reasoned* cataclysm, rather than a chaotic and despondent one. While most chapters in this book are devoted to fundamentally rethinking conventional modern categories such as economy and technology, others are concerned with challenging the excesses of the ontological turn. This approach reflects my conviction that to be profoundly critical of global injustices and the deterioration of the biosphere is not tantamount to jettisoning all faith in science and analysis. We do need an ontological transformation, but for it to serve the interests of justice and sustainability we must retain some of the fundamental conceptual tools and convictions that we have established over the past three centuries. If we had all adopted the tenets of the ontological turn, we would not even have known that the combustion of fossil fuels leads to climate change. We would have found it incomprehensible that someone should suggest that we are now living in the Anthropocene.

Why, then, do I even engage in discussion of posthumanism and the ontological turn? This is not simply because much of anthropology and other social sciences have drifted in that direction, but because some of its proponents have raised crucially important issues that, properly dealt with, may be fundamental to our capacity to challenge business as usual. I am thinking particularly of some central ideas of the French philosopher

---

[6] I believe that the most crucial and fundamental challenge for critical transdisciplinary research on global justice and sustainability is to accomplish radical shifts of perspective through rearrangement of conventional categories and reconceptualization of familiar phenomena. This is not to say that the systematic application of new perspectives to empirical data is any less important, but that a paradigm shift is essential to such work.

Bruno Latour. Although, as I make clear particularly in Chapters 2, 10, and 11, I have often found his style of writing obscure, he has made important points about the social nature of technology and how we tend to purify, as *either* social or natural, phenomena that should not be exclusively assigned to either side. For decades Latour has asked us to acknowledge the formative role not only of social relations in shaping artifacts but also that of artifacts in shaping social relations. This perspective has been fundamental to his contributions to so-called Actor-Network Theory, and to what is now being referred to as an "object-oriented ontology." Although, as I show in Chapter 10, I disagree with Latour's assertion that artifacts can be said to have agency, it is indeed essential to rethink the ways in which the objects around us contribute to shaping our behavior. Even if they do not do so *on purpose*, the specific features of their design have consequences for how we organize social life. It is thus incumbent on social theory to understand how the human fabrication of particular artifacts tends to delegate to *objects* the power of organizing society. As mentioned, this phenomenon is precisely what prompted Marx to apply the concept of "fetishism" to how we let money and commodities rule our lives. Yet Marx's and Latour's approaches to the social life of objects are diametrically opposed. Whereas Marx wanted to expose fetishism as a kind of magical thinking that should be abandoned, Latour appears to propose that we should accept and embrace the influence of objects. Where Marx wanted to show that the apparent agency of objects was illusory, obscuring the unequal social relations that they really represent, Latour dismisses such a stance as condescending and instead asks us to concede that objects have purposes.

The difference between these positions is crucial. Our challenge is to acknowledge the decisive social role of objects without ourselves succumbing to fetishism and attributing purposes to them. The latter would be tantamount to displacing responsibility for social conditions from humans to things, whereas it is essential to realize that the objects that prompt us to behave in certain ways are in themselves *designed* by humans. This means recognizing that responsibility for the organization of society is ultimately our own, but that to transform social relations we must transform the artifacts that compel us to pursue particular kinds of behavior. This is the central message of this book to all those social movements and individuals who, like me, want to see fundamental changes in how world society operates. Its various implications are explored in the chapters that follow. One of them is that efforts to radically transform society will be useless unless we fundamentally redesign the artifacts through which it is reproduced. Another is that it is pointless to blame individual people or an abstractly conceived "system" for the social and ecological evils generated by imperatives emanating from artifacts whose pivotal role as the source of these evils is not even acknowledged. The politics of social transformation are ultimately not about confronting a human enemy but about making collective decisions about transforming the artifacts that make some humans our enemies. A third observation is that it is fruitless to debate economics as if different schools of thought represent alternatives in terms of who is more or less "right" about the most efficient way of handling money, as long as the artifact of general purpose money is axiomatically taken for granted, because the processes of globalization associated with neoliberal economic policies are ultimately inherent in the injunctions generated by that artifact. To advocate curbing such processes without redesigning the artifact that spawns them will create problems of other kinds, but no less severe, than the problems of justice and sustainability that those processes inexorably engender.

The difference between Latour's approach and the perspective presented in this book can be highlighted by the two possible implications of the observation that artifacts have *designs*. Whereas Latour would equate such designs with *purposes*, my objection is that the inclination of artifacts to generate particular kinds of social behavior is inscribed in those artifacts by the humans who have designed them. To be sure, such inscriptions can be very intentional in the sense of being parts of conscious human strategies, but much of the social and ecological ramifications of artifacts are unintended. In any case, it is always incorrect to impute purposes or intentionality to abiotic objects. Artifacts may systematically make people inclined to behave in certain ways, but rather than attribute purposes to the artifacts, we must trace their social consequences to the human activity of designing them. This fundamental premise has emancipatory implications because it rejects magical thinking and allows us to discern ways of transforming society. Although it may seem trivial and obvious when we reflect on the design of most of the technologies with which we interact each day, it presents us with a much more profound challenge when we consider the artifact of general purpose money, which we may not even think of as an artifact, or as something about which we have a choice.

Relations between humans and their artifacts are inherently ambiguous. Approaches such as Actor-Network Theory acknowledge that objects may seem to exert control over humans, but an emancipatory social science must emphasize that artifacts are ultimately controlled by humans because they are designed by them. The argument in this book has undoubtedly been influenced by the insights of object-oriented social science about the causal significance of artifacts, but it moderates this shift of perspective by stripping it of fetishism, anthropomorphism, and misleading injunctions to dissolve distinctions between subjects and objects and between nature and society. In these respects, I want to hold on to the Enlightenment worldview that was also the point of departure for Karl Marx's analysis of capitalism.

I finished the final note in my previous book, *Global Magic*, with the conclusion that, "if we fail to see the problem with money, we fail to see the problem with technology." That conclusion, in a nutshell, summarizes a central point of departure for this book. To grasp, as succinctly as possible, the conundrum of the money-energy-technology complex, we can briefly return to the case of energy policy in Cuba. For decades, Cuba has aspired to shift to renewable energy technologies such as photovoltaic power, but its electricity has remained almost completely dependent on fossil energy. The reason why the country has not shifted to solar power is profoundly ironic: explanations generally refer to a lack of funding. Applying a Marxist idiom that should be familiar to Cuban politicians, this is tantamount to saying that the progress of the "productive forces" in Cuba has been constrained by a lack of *capital*. But capital, we learned from Marx, is the product of processes of accumulation based on inequitable social exchange relations and the use of industrial technologies propelled by fossil fuels. Is such unethical and unsustainable accumulation of capital a requisite for technological progress even when we are considering a shift to renewable energy? As I argue in Chapter 7, my answer is yes, but it runs counter to how technology is perceived in both mainstream and Marxist accounts. In both orthodox and heterodox economic thought, the ontology of technological progress is conceptually sequestered from the organization of societal exchange. Our trust in the efficacy of technology as a phenomenon excised from global resource flows – what I have called *machine fetishism* – is inextricably connected to how mainstream and Marxist economics tend to ignore the material asymmetries of ecologically unequal exchange. As I show in Chapter 6, the complete neglect of global, distributive aspects of technologies, as revealed by their dependence on asymmetric flows of biophysical resources, including embodied labor, is a remarkable feature of the otherwise multifarious field known as the "philosophy of technology." After more than two centuries of rapid technological change, philosophers of technology appear not to have seriously considered the possibility that modern technology is not so much a means of saving as of *displacing* demands on human time and natural space onto other populations with less purchasing power. This neglect of the sociometabolic dimension of technology illustrates the kind of fragmentation of perspectives that we must transcend to grasp our escalating problems of global inequality and unsustainability. To comprehend the logic of the money-energy-technology complex, we must integrate critical perspectives on money with theories of ecologically unequal exchange and their relation to the phenomenon of technological fetishism. As I elaborate in Chapter 7, we need to recognize how even the technologically defined relation between time and space, represented as transport velocity, implicates strategies of appropriation.

The conceptual sequestration of engineering vis-à-vis economics is a fundamental dilemma of modernity. It has prompted me to challenge the ontological status of technology not only in mainstream thought, as in Chapter 6, but in Marxist theory as well. Although I am obviously strongly influenced by Marxist concerns with fetishism, exploitation, and justice, I do not share the value theory or the understanding of technological progress that are prevalent in Marxist discourse. In my view, these different components of a Marxian worldview must be disentangled from each other for the full significance of the concepts of fetishism and exploitation to protrude. This is my objective in Chapters 8 and 9.

I hope to have made clear why, to conceptualize the money-energy-technology complex and its relation to the Anthropocene, we must engage with questions of ontology, fetishism, and the ways in which we conventionally distinguish between nature and society. This in turn requires consideration of perspectives from a diverse set of approaches ranging from Marxism to posthumanism. To be able to accept the fundamental reassessment of money and technology that I am proposing in this book, we must be prepared to seriously consider not only conventional approaches of economics and the heterodox challenges to them but also the radically alternative critiques that seek to deconstruct and undermine the entire rationalist framework in terms of which mainstream deliberations on sustainability and inequality are being conducted. This in turn demands careful assessment of the different ontological assumptions that underlie neoliberal economics and engineering, Marxist critiques, and posthuman deconstructions of the allegedly Eurocentric, colonial premises of modern thought and discourse. In Chapters 10, 11, and 12,

I explore some posthumanist challenges to the tenets of mainstream science, primarily to determine the limited extent to which a critique of Anthropocene extractivism necessarily implies a complete rejection of those tenets.

As I argue in Chapter 12, we can transcend our fetishized ways of reifying "money," "energy," and "technology" by recognizing these concepts as referring to *relations* rather than things. All three are relational phenomena, although they are conventionally perceived as self-evident objects that can be defined as if they have an autonomous essence. The argument in Chapter 7 is that energy denotes a relation between humans and their environment that is defined by the state of their technology, which is in turn contingent on the amount of money available to them, which itself is an index of their relation to other people. This is the essence of the money-energy-technology complex. It is what defines modernity as a social condition founded on the capacity to externalize biophysical burdens and risks (not to be confused with "costs") and to outsource demands on resources and labor. This capacity creates illusions of cornucopia, as if there were no ecological limits, but modernity only *displaces* limits, rather than dissolves them. Imperialism has always involved strategies of time-space appropriation (e.g., through tribute of the products of labor and land), but modernity is founded on a new such strategy – the money-energy-technology complex – which still has to be recognized as a variant of imperialism.

If market globalization and technological intensification are to be understood as sophisticated strategies of appropriation through displacement, the obvious antidote must ultimately be a return to the autonomous carrying capacity of the land we inhabit. By "land" I do not mean the artificially overproducing fields of an industrial agriculture running on fossil energy and nourished by ghost acreages on other continents, but a piece of the Earth's surface capable of providing sustainable returns on human labor without depriving other people and other generations of their time and space. Such a vision of humanity's future is infinitely distant from what currently appears realistic, viewed from modern urban lifeworlds founded on seemingly cornucopian supermarkets. We need to remind ourselves, however, that the alternative that we increasingly view as realistic is nothing less than an "uninhabitable" planet (Wallace-Wells 2017).

It is noteworthy that Bruno Latour's most recent book *Down to Earth* (Latour 2018) appears to advocate such a reorientation toward learning to inhabit our local territory. After decades of convoluted philosophical detours, posthumanism may thus be converging with the message reiterated by generations of back-to-the-landers since the early nineteenth century. Has it suddenly discovered the countless marginalized voices endorsing the frugality of peasants, indigenous groups, bioregionalists, and other "ecosystem people"? If so, there is a voluminous and time-honored discourse to draw on, expounding the virtues of attachment to place, traditional ecological knowledge, deep ecology, self-sufficiency, reduced ecological footprints, degrowth, localization, and a myriad similar, decidedly nonmodern ideals. As usual, however, posthumanism offers us perspectives that are more evocative than analytical. I criticize this diffusive inclination of posthumanist discourse particularly in Chapter 11. To argue for a nonmodern relation to land, we must first rigorously unravel how the money-energy-technology complex defines the modern condition. This is not something that can be accomplished through the elusive jargon of posthumanism. While my conclusion largely converges with that of the posthumanists, the argument is founded on the kind of analysis that they reject.

Like the posthumanists, I ask the reader not to take anything for granted – neither the money nor the technologies by means of which we all conduct our everyday affairs. They are not the self-evident and politically neutral components of modern life that we tend to think. They all rely on asymmetric ratios of resource exchange on the world market, and on processes of energy conversion that are destroying the biosphere and the conditions for human life on this planet. We have been aware of the resulting inequalities and environmental problems for generations, but the mainstream political, economic, and technological proposals for mitigating them have generally been counterproductive. Challenges to business as usual have assumed that there is a group of people to blame – such as capitalists, economists, or foreigners – but the bottom line of this book is that power resides in the artifacts that rule our lives, *until* we discover that this is so, that we are the authors of those artifacts, and that they can be redesigned.

Rethinking Economy and Technology

I have recently been involved in debates with posthumanists about how to challenge the economic, political, and ecological absurdities of our contemporary world. It is easy for most of us to agree that the current trajectory of global society is a source of frustration, to say the least, and I don't think I need to refer to statistics on rising inequality or the transgression of "planetary boundaries" to amplify that feeling. A significant number of anthropologists and other social scientists today seem to want to turn their backs on modernity. For many anthropologists, this is an *ontological* turn. It means identifying with the nonmodern worldviews of the people who have hosted them during fieldwork. Some problems with this approach are discussed in Chapters 10, 11, and 12 of this book.

There was a time when I, too, toyed with the idea that animism could show us a way out of modernity. But my position for many years now is that there is no way to resurrect ontologies that have proven incapable of resisting modernity. Our only hope is to understand, in new ways, what modernity *is*. And – no matter how profoundly enlightening it may be to immerse oneself in perspectives beyond the modern worldview – this is something that I think we shall never learn from our nonmodern Others, who tend to become as attracted by, dependent on, and ultimately possessed by modernity as we are. We need to understand the cultural, economic, and ecological logic of the global processes that have brought us to this point – the so-called Anthropocene. As I argue in Chapters 2 and 11, the urgent challenges of the Anthropocene cry out for our serious attention, beyond all the imaginative ideas that preoccupy the posthumanists and the proponents of an ontological turn. This does not mean immersing ourselves in mathematics. As I hope to show, there is a space between algebra and poetry, where language can be used transparently to communicate about the subtle interpenetration of factors deriving from features of nature and features of society. (And yes, as you will see, I suggest that we retain this distinction.)

To understand what modernity "is," let us consider an early-nineteenth-century coincidence that I do not think *was* a coincidence. I am thinking of the simultaneous birth, in the very same place, of the idea and phenomenon of "technology" and the idea and phenomenon of "economics." You will probably object that technology and economics existed long before the early nineteenth century – and perhaps you will remind me of medieval watermills or even Paleolithic stone axes, and of the preindustrial economic doctrines of the mercantilists or the Physiocrats. However, by technology and economics I mean things and ideas that are contingent on a globalized market, like the steam engine used for producing cotton textiles for the Atlantic slave trade, and like David Ricardo's concept of "comparative advantage." There was a definite historical discontinuity, in the early decades of the nineteenth century, in terms of how technology and economics were perceived. We might even talk about the "invention" of these two fields as autonomous domains of thought and practice. The Industrial Revolution gave us a new concept of technology, and classical economists such as Ricardo developed the basic vocabulary of economics. Both these developments occurred in the core of a world empire of unprecedented scope and power. This, I maintain, was not a coincidence.

What we need to understand – to stand a chance of keeping the Earth inhabitable for our great-grandchildren – is the relation between these three phenomena: technology, economics, and imperialism. What is the relation between the emergence of the disembedded "economy" and the emergence of a new and disembedded "technological" rationality, and how is the emergence of these new *categories* related to core–periphery relations in the world-system? And to address our current predicament: How are these questions implicated in contemporary debates about sustainability, global warming, and the Anthropocene? Even more fundamentally, how are they related to the issue of how we distinguish between nature and society?

In focusing on the operation of markets, the so-called neoclassical school of economics that was established in the 1870s definitively abandoned the concerns of some earlier economists with the material substance and the history of production of traded commodities. From now on, mainstream economics was to be exclusively concerned with how much money commodities fetch on the market, rather than with how much labor,

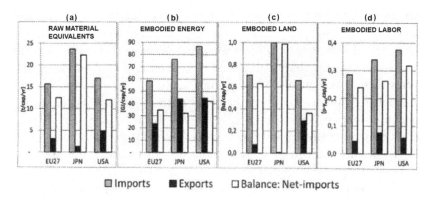

FIGURE 1 Per capita net imports of biophysical resources to the European Union, Japan, and United States in 2007 (diagrams compiled by C. Dorninger)

land, energy, or materials had been spent producing them. While earlier schools had been concerned with the relation between economic value and material inputs in the production process – such as labor or land – these topics were completely abandoned with the Marginalist Revolution. The effect was a final detachment of economics from considerations of the biophysical metabolism of human societies. This detachment has been the source of a long-standing critique from heterodox economic perspectives such as those of Marxism and ecological economics (Martinez-Alier 1987). However, the economic perspectives of Alfred Marshall and Karl Marx – although diametrically opposed – can both be traced to Ricardo. The former focused on his modeling of market exchange, the latter on his conviction that the source of exchange-value was human labor.

What can we say about the biophysical metabolism of world society in our own time? Statistics on modern world trade converted from monetary to biophysical metrics reveal *net* imports to the wealthiest core areas of the world-system – the United States, the European Union, and Japan – in the year 2007 totaling around 12.6 gigatons of raw material equivalents, 34 exajoules of embodied energy, 5.6 million square kilometers of embodied land, and 247 million person-year equivalents of embodied labor (Dorninger and Hornborg 2015; figure 1).[1] As all three areas were

---

[1] The diagrams in Figure 1 show net imports per capita. They reveal, for instance, that an *average* American, one-child household in 2007 had the equivalent of one full-time servant outside the US border, and that such a household in Japan consumed the produce of three hectares of land outside the country. It follows that an average American household with

net importers of all four resources, the statistics show that it was not simply a matter of specialization, with different areas exchanging different kinds of resources with each other, but instead a net transfer of all four resources contributing to the accumulation of global technological infrastructure as a whole. This incontrovertible, material asymmetry of the global economy appears to be of no concern whatsoever for mainstream economists. Two hundred years after Ricardo, they continue to assert that free trade is good for everyone.

The point I want to make is that the shift of theoretical perspective toward an increasing preoccupation with market equilibrium was not simply a disciplinary development occurring in a political and ideological vacuum but instead a discursive rationalization of the material asymmetries of colonialism. It permitted the asymmetric resource flows of the British Empire to continue, invisibly, beyond the official end of colonialism. To this day, the vocabulary and theoretical assumptions of nineteenth-century economic thought continue to legitimize the operation of the world market, even as increasing numbers of people are alarmed by its tendency to widen global inequalities and – no less significantly – to degrade the ecological conditions for human life. In this book I hope to suggest the contours of a crucial ontological shift in our understanding of economy and technology, which might increase our chances of surviving the Anthropocene. To achieve such a shift, we need to pursue a middle road between constructivism and realism. We shall need the ethnographer's readiness to deconstruct and defamiliarize his or her own categories, but also the scientist's conviction that the world in which we are immersed has objective properties that can be more or less successfully represented. To establish such a middle ground between relativism and objectivism, we must acknowledge analytical distinctions between the natural and the social, the material and the semiotic, and the objective and the subjective.

The title of my previous book is *Global Magic*. I want to explain how the topic of magic is relevant to these critical perspectives on economic history, sustainability, and global justice. There is a vast literature on this topic in anthropology, but out of a great number of possible contributions I shall just mention one, because it highlights, as well as any other, the essence of what we modern people tend to classify as magic. Drawing on the ethnology of colonial Indonesia, Margaret Wiener (2013) has shown

four children in that year had *two* people abroad working full-time to cater to its needs, and that its Japanese counterpart required *six* hectares overseas.

that a criterion used by modern Europeans to dismiss something as magic is that its efficacy is *contingent on human consciousness*. The Indonesian practice of *guna-guna* poisoning, once it was understood by the Dutch colonists as materially and physiologically harmless, could be dismissed as "superstition" in the sense that it was dangerous only to people who *believed* in its magic. Referring to Bruno Latour, Wiener confirms that modern Europeans tend to emphasize a divide between the objective, material causality of nature and the cultural and psychological efficacy generated by society. Stanley Tambiah (1990: 108–109) similarly distinguishes between notions of "causality" and "participation." While the former is inexorable and incontrovertible, the latter is contingent, constructed, and open to negotiation. This is why so much of the anthropological discussions on how to delineate magic has focused on its rhetorical and performative character – in other words, its capacity to *persuade* (ibid.: 80–82).

Against the background of this basic distinction, what is classified as nature and what is classified as society has very significant ideological implications. Phenomena classified as nature or natural are automatically understood by modern people as intrinsically uncontaminated by social or cultural processes. Although nature can be conquered and modified, it remains *conceptually* impenetrable, and consequently exempt from critique. While it is important to retain the idea of nature as *analytically* distinct from society, a very real problem is that the "socionatural" phenomenon of technology has been classified as belonging *exclusively* to nature.

Wiener's constructivist perspectives on magic – and the questions she asks – are somewhat different from mine, but her identification of the essence of a modern definition of magic confirms my own use of the concept. I must admit to sharing the modern and currently contested conviction that the distinction between subjective and objective aspects of causality *is* significant. To exemplify, I have pointed to the difference, in this respect, between keys and coins. Both these little pieces of metal can be attributed with agency in the sense that they can open doors, but keys do so because of their physical shape, while the ability of coins to do so is contingent on human consciousness: namely, the beliefs of the doorkeeper about their value. In conventional modern thought, then, keys could be classified as technological artifacts, while coins could be classified as magical. Indeed, Marx referred to our belief in money as "money fetishism," derived from a Portuguese word for sorcery (*feitico*). The important thing to acknowledge here is that the delegation of agency to artifacts can be dependent on, or independent of, human consciousness.

When I refer to our "belief" in money, I mean it in both senses unearthed by Malcolm Ruel ([1982] 2002) and Jean Pouillon ([1982] 2016): the propositional belief *that* money has value, and the trusting belief *in* money as a solution to many of our problems. Both nuances of the concept of "belief" refer to aspects of human consciousness. As we shall see, the same semantic duality tends to apply to the modern belief in technology: the proposition that advanced technology is a politically and morally neutral resource potentially available for all humankind, and the trust that it will solve our various problems. Significantly, the concept of economic "credit" is closely intertwined with trust and the expectation of reciprocity (Pouillon [1982] 2016: 487; Bourdieu 1991: 192). Money has extended the trust in reciprocity beyond the field of personal relations, and concomitantly widened the scope of unequal exchange, which, defined in biophysical terms, reveals the illusory character of abstract market reciprocity.[2]

The distinction between agencies that are dependent on, or independent of, human consciousness is one that we are not likely to encounter in Actor-Network Theory, the ontological turn, or posthumanism in general because this distinction presupposes a contrast between the subjective and the objective, which posthumanists tend to reject. And yet, I would argue, such a distinction is crucial to achieving the kind of deconstruction of the global techno-industrial system that many posthumanists would endorse. As I suggested in *Global Magic*, "[R]ather than championing a magical ontology that most of us have irrevocably lost, an anthropological approach is more usefully applied to exposing the unacknowledged magic of our *own* ontology" (Hornborg 2016: 111; emphasis added). One of the central points of that book is that technology – the seemingly objective operation of increasingly complex physical artifacts – is contingent on economics – the subjectively constituted, socially constructed ratios by which the various components of technologies are made accessible to different social groups. This is what I mean by "machine fetishism." The appearance of solid, material objectivity obscures the arbitrary social exchange relations that make the machine possible – for those who can afford it.

---

[2] It is also significant that the link between trust and expectations of reciprocity is foundational to the phenomenon of sacrifice (Pouillon [1982] 2016: 491), which suggests a profound connection between economics and religion.

Latour and other constructivists may be finding themselves in a bit of a quandary, now that it would appear impossible to relativize the alarming facts of the Anthropocene as "social constructions" or as what Latour (2010) calls "factishes." Could proponents of the ontological turn seriously assert that the threat of climate change represents the particular perspective of Earth System scientists? I do not agree with Latour that we should abandon the categories of nature and society – no more than we should abandon the categories of object and subject.[3] However, an important point that Latour has made is that there is a modern tendency to "purify" socionatural phenomena as belonging to either nature or society. This observation might help us approach current concerns with sustainability as well as my own argument on machine fetishism. Since the Industrial Revolution, mainstream economists have believed that their models can account for economic progress without any consideration of nature. For two centuries, in other words, the market has been sequestered from nature. It is conceived as a purified social phenomenon. No wonder the economists are now unable to deal with the incapacity of modern society to stay within planetary boundaries. Climate change is one "externality" that I don't think even economists seriously believe can be internalized in market prices.

My argument on machine fetishism is the mirror image of this unwarranted purification of the economy. Since the Industrial Revolution, engineers have believed – like the rest of us – that technological progress can be understood without consideration of the structure of world society. For two centuries, the machine has been sequestered from society. It is conceived as revealed nature. No wonder technology continues to be very unevenly distributed in the world-system.

These one-dimensional perspectives on what we should understand as socionatural processes of exchange and production are a consequence of how our thinking about economy and technology has been compartmentalized since the nineteenth century. The point of departure of this sequestration of economics from nature and technology from world society is that the natural components of technologies are translated into market prices that have little to do with their material substance. These market prices – representing societal exchange rates – determine the feasibility of

---

[3] But note that the two distinctions do not coincide. As I argue in Chapter 10, there are social objects and natural subjects, and although the posthumanists tend to approach the subject–object distinction as simply political, it is ultimately an ontological one: a cat may *treat* a mouse as an object but ontologically the mouse *is* undeniably a subject.

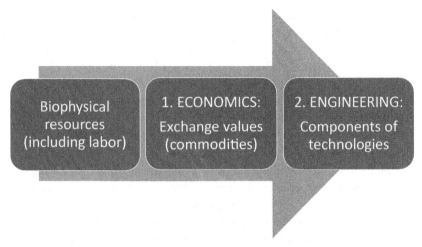

FIGURE 2 The two-step purification of asymmetric resource flows into illusions of
(1) market reciprocity and (2) the technological revelation of (nonsocial) nature

a given technology, yet that technology is understood primarily as the
revelation of intrinsic features of nature. Thus, invisibly and insidiously,
biophysical resources are first asymmetrically exchanged within a highly
unequal world society and then unevenly assembled into technologies, as
if the technologies were simply indices of the ingenious exploitation of
nature, through advanced engineering (Figure 2). This is the illusion of
technological progress that has distorted our understanding of human–
environmental relations since the Industrial Revolution.

I guess I should explain how I can find Latour's observations on
purification helpful and significant while I am unpersuaded by his injunction to abandon the distinction between nature and society. Am
I contradicting myself? I think not. To say that nature and society are
intertwined in our bodies, our landscapes, our technologies – and in
climate change – is not to say that there is no such thing as natural versus
societal *aspects* of these phenomena. Analytical distinction is not the same
thing as rigid ontological dualism. Also, Latour's many examples of the
misleading implications of purification are all confined to concrete artifacts, specific technologies, and particular objects of scientific research,
but the most compelling critique of purification is achieved by applying it
at the abstract level of modern categories such as "economy" and "technology." In other words, it is not the acknowledgment of the categories of
nature and society that is problematic, but the wholesale classification of
entire fields of inquiry as belonging to one or other of these domains:

referring the study of exchange to an exclusively social domain, and the study of production to an exclusively natural one.

Much has been written on how the establishment or invention of the categories of "economy" and "technology" was part and parcel of the Industrial Revolution. From different vantage points, for instance, Karl Polanyi ([1944] 1957) and Robert Heilbroner ([1953] 1999) have discussed the emergence of the disembedded market economy as simultaneously a topic of discourse and a social practice. Parallel to this disembedding of economic rationality, there emerged in early industrial Europe a new and disembedded technological rationality. Although there have been voluminous deliberations, over the past two centuries, on the societal, cultural, political, and ecological consequences of economic and technological change, through all this discourse there has remained an almost self-evident separation of economy and technology as distinct phenomena. This separation appears to reflect the obvious fact that economists and engineers deal with quite different things.

The separation of economy and technology – concerns with flows of money and with the harnessing and reorganization of matter and energy, respectively – does indeed suggest a misguided distinction between the social and the natural. Marx in the mid-nineteenth century recognized that flows of socially constructed exchange-values and of biophysical resources such as human labor-power should be integrated into a single theoretical framework to reveal how increasing social inequalities were generated by the regular operation of the market economy. As argued also by a long line of early proponents of ecological economics (Martinez-Alier 1987), the new science of economics was too preoccupied with monetary flows and market equilibrium to acknowledge the physical dimension of human societies. Both these categories of heterodox economists thus challenged the inclination of mainstream economists to disregard the material metabolism of society. Their efforts have converged in what is currently referred to as "ecological Marxism." In Chapters 8 and 9, I discuss how this acknowledgment of the material aspects of the economy should prompt us to rethink the Marxian concept of "value."

To account for the simultaneous historical emergence of modern economic and technological rationality, I propose that the invisibility of nature in mainstream economic discourse is conducive to the invisibility of world society in mainstream engineering science. The exclusion of biophysical aspects from economics is the basis for the modern kind of technological rationality, for which considerations of material frugality are simply superfluous. What Aristotle called *chrematistics* – the management of money – has made economics in its original sense – that is, economizing with substantive resources – obsolete. For instance, the economists' cost–benefit analyses do not reckon with biophysical resources such as energy, only with relative monetary prices. This is why concerns with net energy, or EROEI (Energy Returned On Energy Investment), remain irrelevant for neoclassical economists, and why the notion of "technological progress" for economists and engineers alike is quite compatible with diminishing returns on energy and other resources.

This argument is not a conspiracy theory imputing malicious intentions to economists or engineers. It is an account of how the sequestration of nature from economics and of society from engineering has been able to maintain the illusion that technological systems are not founded on appropriation. Once the market is perceived as an arena for the flow of abstract exchange-values, and the potential of engineering is perceived as defined by this market, the construction of labor-saving devices becomes simply a morally and politically neutral project of harnessing nature's sources of energy for the benefit of humankind. The fact that the inputs of resources into labor-saving devices are made feasible by a given constellation of market prices – that is, terms of trade – is sequestered from the technicalities of engineering, so that engineering appears to its practitioners as completely detached from the global organization of society. Not only are material asymmetries in exchange thus made invisible by the ostensibly neutral operation of market mechanisms, so is their significance for the feasibility of specific technological systems that rely on inputs purchased on the market. The crucial point I want to make is that modern, globalized technologies represent not simply politically neutral revelations of possibilities inherent in nature, but socionatural machinations that are *contingent* on asymmetric transfers of resources in world society. In other words, modern technologies are products not just of engineering – viewed as innocent research into the physical regularities of the material world – but finally also of unequal societal relations of exchange.

For almost a hundred years, anthropologists have been intrigued by Marcel Mauss's ([1925] 2016: 73) observation that gift giving among the premodern Maori of New Zealand formed ties between human souls because the gift was believed to have a soul that could "create a magical or religious hold over you." One of the questions I ask in *Global Magic* is what it means to say that modern people have *dismissed* this belief in the magical agency of objects. The distinction between the gift and the commodity has been a topic of discussion for generations of

anthropologists. Whereas the magic of the gift is founded on the conflation of people and things, the modern commodity is instead a result of
the alienation of people and the products of their labor. In the former
case, things are represented as persons; in the latter, as Marx observed,
relations between people assume the form of relations between things.
Paradoxically, however, *both* conditions qualify as forms of fetishism.
Whether perceived as metonymical extensions of their givers, or as
autonomous embodiments of abstract labor, the things with which we
surround ourselves thus magically implicate social relations.

At this point, my recurrent references to the concept of fetishism no
doubt demand a definition. I use the concept to denote the attribution to
an object of properties that are assumed to be – or treated as if they were –
inherent in its internal constitution, but that are in fact contingent on
social relations. In the idiom of C. S. Peirce's semiotics, Charles de
Brosses's argument (Morris and Leonard 2017) was that fetishes are
indices rather than symbols, whereas Marx's insight was that money
and commodities are *incorrectly understood* as indices, although they
signify social relations of exchange. Marx's contribution was to show
how our preoccupation with the manipulation of objects underpins relations of exploitation by deflecting our attention from those relations. To
say that social relations are fetishized is to say that they are reified into
objects that in turn become vehicles of social games – money for the
economy, commodities for consumption, and technology for engineering –
in which the participants tend not to be cognizant of the societal requisites
of the existence of those objects. Social games contingent on fetishization
thus deflect attention from potentially provocative conditions such as
inequalities, exploitation, and power, focusing instead on the ostensibly
neutral manipulation of seemingly innocent "things." This is not a matter
of conscious conspiracy but of a pervasive human inclination to perceive
artifacts as excised from the social relations that they represent.

A question we might ask ourselves is if premodern anxieties about
failed reciprocity are less justified than the belief that all concerns with
reciprocity can be delegated to the market. For this is precisely the
ideological function of the conviction that market prices guarantee reciprocity: the anonymous and seemingly incontrovertible exchange rates on
the market are supposed to liberate us from any moral qualms about the
distant social and ecological repercussions of our transactions. As the
Dark Mountain Manifesto (Hine and Kingsnorth 2009) puts it, the story
of our civilization is "the story of a people who believed, for a long time,
that their actions did not have consequences." In disembedding both

exchange and production – "economics" and "technology" – from face-to-face social relations and close contact with the sources of human sustenance, modern society gives our modern lifestyles an illusory aura of moral and political neutrality. The abandonment of "superstitious" beliefs in the magical agency of unreciprocated gifts has been perceived as progress, rationality, and enlightenment, but this new perception of commodities as detached from the social contexts of their production is conducive to another *kind* of magic. Modern magic is not about fearing personal retribution through an intentional agency attributed to objects, but about ignoring the increasing extent to which our lifestyles have deleterious impacts on other people and global ecology, and the increasing likelihood of a major backlash. Whereas premodern magic was the source of an uncanny fear of personal, purposeful evil, modern magic imposes an impersonal evil, generated by our lifestyles, afflicting people and ecological contexts far beyond our horizons – and likely to generate an intangible, disembedded, and impersonal retaliation. We must finally ask ourselves why it should be considered more "rational" and "enlightened" to pretend that our market transactions have no moral implications, than to allow our exchanges to burden our conscience, as among the premodern Maori.

In masquerading as reciprocity, the exchange ratios defined by the world market continue to direct net transfers of resources to core areas of the world-system. And in masquerading as revealed nature, modern technologies can continue to globally appropriate human time and natural space as if they were not contingent on unequal exchange. At the core of modern magic is the purification of the machine. Since the Industrial Revolution, economic growth and technological progress have served as supremely efficacious strategies for displacing workloads and environmental burdens onto other people and other landscapes. Viewed as strategies to achieve such displacement, they belong to a category of societal arrangements that includes slavery and imperialism. If we approach modern, globalized technologies – that is, technologies that are contingent on market prices – as strategies of appropriation, we might conversely view premodern strategies of appropriation as technologies. Used in this sense, in other words, the concept of "technology" encompasses aspects of social organization as well as physical energy conversion.

One reason why the mainstream notion of technological progress has been so misleading is that the Industrial Revolution – and the very phenomenon of modern technology – can be perceived at vastly different geographical scales. Anthropologists tend to be quite familiar with the fact that the world-system looks different when viewed from a village in sub-Saharan Africa than when photographed by a satellite. The bird's-eye view afforded by a global satellite image of nighttime lights is more conducive to a gestalt shift in our view of the role of technology in the world-system than any amount of time spent interviewing engineers, stockbrokers, politicians, factory workers, or African villagers. The logic of total global dynamics is rarely apparent at the local level. A classic example is the cargo cults in early-twentieth-century Melanesia, which to many anthropologists exemplified how local perceptions tend to be constrained by limited horizons. If we turn this insight back on European history, we might see one reason why the European understanding of technological progress in the early nineteenth century was so detached from considerations of American slave plantations and the deindustrialization of India. To this day, historians of technology tend to focus on the *local* conditions and repercussions of mechanization, whether British deliberations on the so-called machinery question (Berg 1980), the fury of the machine-breaking Luddites (Sale 1995), or capitalist incentives to mechanize (Malm 2016). It is rare to find historians approaching technological progress as a local manifestation of a *global* process, although significant contributions in this direction have been made by some of the "new world historians" like Andre Gunder Frank (1998), Jim Blaut (2000), Kenneth Pomeranz (2000), and Joseph Inikori (2002).

A more general reason why the total logic of what I call the money-energy-technology complex has escaped us is the fragmentation of disciplinary perspectives. The economists' discussions about money tend to be completely detached from the physicists' discussions of energy and the engineers' discussions about technology. As a consequence, empirical revelations about ecologically unequal exchange or the political rationale of energy transitions are not permitted to contaminate our image of the nature of modern technology. Conversely, as I emphasize in Chapter 6, constructivist insights on so-called sociotechnical systems (Bijker, Hughes, and Pinch 1987) are not in the least concerned with global metabolic asymmetries, or with the perspectives of the new world historians on the accumulation of industrial technology in nineteenth-century Britain. For instance, most proponents of the material turn and the "new materiality" in the humanities and social sciences have probably never heard of the quantitative methods of material flow analysis. Their focus on the local phenomenology of material objects is completely detached from the alarming reports of the UN Environmental Programme (Schandl et al. 2016), which showed that, rather than "dematerializing," the material intensity of the global economy has been increasing over the past decade.

The historians and philosophers of technology and sociotechnical systems seem completely uninterested in money, energy, and material flows, yet the biophysical metabolism of the world-system is precisely a sociotechnical system. Like the blind men trying to grasp the features of an elephant, conventional academic disciplines are unable to grasp the logic by which the categories that organize our production and relations of exchange are propelling us toward what David Wallace-Wells (2017) called an "uninhabitable" Earth. There is indeed an elephant in the room, and it is called money – the seemingly natural idea that everything on Earth is interchangeable: rainforests, Coca-Cola, coral reefs, iPhones, human lives, and video games. Although our global predicament is being blamed on every conceivable *consequence* of the operation of money – such as modernity, capitalism, globalization, growth, technological progress, or even economic theory – critics of each of these various aspects of that predicament rarely trace its roots to the idea of money as currently designed. Yet that is precisely where a critical anthropology of the Anthropocene might begin. As I argue in Chapters 2 and 11, a central paradox of the Anthropocene is that to some it suggests a rejection of "human exceptionalism" – as if humans were just a species among other species – while underscoring our inevitable anthropocentrism. No other species could have invented money. No other species could be transforming the planet as rapidly as we are. No other species could name a geological epoch after itself.

The predicament of the Anthropocene generates different existential coping strategies among different people. If we think of our existential options as arranged along a spectrum, the opposite end points would be Donald Trump's denial of climate change versus the creative resignation of the Dark Mountain Manifesto (Hine and Kingsnorth 2009). To the former, there is no problem, while to the latter, there is no solution. In between is the ecomodernist optimism of the Breakthrough Institute (toward Trump's end of the spectrum) and the critical activism of the climate justice movement (toward the other end). Although politically opposed, both these latter positions agree on the reality of climate change and the feasibility of technological solutions. The main difference between them is whether they think such solutions require radical societal change. And in the middle, finally, are the vast majority of human beings who must simply repress the whole issue to be able to go on with their lives, conducting business as usual. Common to most people worrying about climate change, except the position exemplified by Dark Mountain, is a faith in technological solutions. History will tell whether it is the destiny of our species to march – like lemmings – quickly and blindly into oblivion, or whether it will be able to muster the kind of responsibility and restraint that would be required for it to consciously redesign the very idea that has made it master of the planet. I refer, of course, to money. That appears to be our only hope. In Chapter 13, I venture some suggestions on how this might be done.

One of the paradoxes of our way of writing history is that we do not hesitate to account for the accomplishments of earlier civilizations – ancient China, Egypt, or Rome – in terms of exploitative class relations and the appropriation of resources by a powerful minority, yet we tend to tell ourselves that the accomplishments of modern industrial society over the past 250 years are *not* products of exploitation. We use quite neutral and benevolent words to describe these accomplishments: "modernization," "economic growth," "industrialization," "development," "technological progress," and so on. Yet the increasingly obscene inequalities of the modern world far surpass those generated by earlier civilizations. As the anthropologist Maurice Godelier (1994: 105) has proposed, it is incumbent on us to grasp how a given social system is able to represent unequal exchange as reciprocal. It is with this imperative in mind that we have every reason to scrutinize the cultural and ideological logic of general purpose money and globalized markets – and its manifestations in unevenly distributed machinery.

In experiencing the globalized technological infrastructures of urban life in the twenty-first century, the individual person is likely to conceive of them as the all-powerful and immutable substance of human society, enormous in its extension and relentless in its expansion. So completely does it dwarf the individual human subject as to make any general critique of modern urbanism seem silly, verging on the psychotic. The sheer volume of resources mobilized in the metabolism of this globalized social organism is inconceivable for the individual preoccupied with managing his or her personal household concerns. Like the architecture of previous civilizations – the amphitheaters of Rome or the pyramids of Egypt – the physical infrastructure of the modern metropole is intimidatingly greater and grander than a person's capacity to encompass it. Its complexity and monumentality cannot but command submission. Yet this unimaginably massive metabolism hinges on the continuous throughput of flows of energy and materials that are ultimately quite vulnerable to disruption. So-called oil crises and electric blackouts are ubiquitously understood as temporary interruptions in the inexorable progress of high-tech society, rather than sources of serious doubts about its long-term resilience. Popular confidence in modern infrastructure is as essential to its operation as trust in the market economy with which it is intertwined.

Let me recapitulate the basic outline of my position. Rather than being intrinsically and generally flawed, the problem with the nature–society distinction is its skewed way of organizing our thinking regarding economics and technology. The fields of economics and engineering are ontological mirror images of each other. Mainstream neoclassical or marginalist economics tends to assume that a theoretical account of economic growth does not need to consider nature, while mainstream engineering tends to hold that a theoretical account of technological progress does not need to consider global society. Both these foundational modern assumptions are false, but I emphasize again that this is not to say that it is misleading to *analytically* distinguish between nature and society. What these assumptions accomplish, however, is to allow us to disregard the fact that our lifestyles and consumption patterns have very tangible consequences for other people and ecosystems, and to pretend that our way of living has no problematic moral or political implications.

Moreover, the purification of economics as social and technology as natural has had implications for our relative readiness to subject them to critique. Economic organization, such as the market, has been open to serious political contestation throughout history, whereas critiques of technology, such as those of the machine-breaking Luddites or – later – the dystopian philosophies of thinkers like Lewis Mumford, Martin Heidegger, or Jacques Ellul, have been dismissed as romantic, regressive, or even ridiculous. Kirkpatrick Sale (1995) tellingly titled his book on the early-nineteenth-century machine-breakers *Rebels against the Future: The Luddites and Their War on the Industrial Revolution*. The message is clear: to question technological progress is to try to halt something inevitable. The widespread leftist vision of a future, postcapitalist society is based on our readiness to reorganize the economy, but as I argue in Chapters 8 and 9, that vision continues to rely on an unshakeable faith in technology. There is little awareness of the extent to which the two are inextricably intertwined.

Of what relevance to these deliberations is the concept of magic? Whereas the kind of magic considered by Wiener (2013) highlights how social forces can penetrate natural causality through self-suggestion affecting the human body, the modern magic that I have tried to analyze hinges on how social organization can penetrate natural causality by means of technology. The former refers to how social relations may intervene in our biological metabolism, the latter to how money fetishism has generated a fundamental transformation in global social metabolism. Pervading the solid material infrastructures of modern urban life is the market's ever-present language of persuasion.

In this book I hope to show that both markets and machines are ultimately founded on human *beliefs*, in the double sense of propositions and faith (Ruel [1982] 2002: 103). The Dark Mountain Manifesto (Hine and Kingsnorth 2009) observes that civilizations are basically sets of beliefs. It cites Joseph Conrad's and Bertrand Russell's insights that human civilization "is built on little more than belief," and that the collapse of a civilization is more or less synonymous with the collapse of a belief system. Economists would agree about how crucial it is that we continue to believe in the magic of money. The financial meltdown on Wall Street in 2008 opened a crack in our belief system. The Dark Mountain Manifesto was written a few months later. From the slopes of the Dark Mountain, it imagines, "[W]e shall look back upon the pinprick lights of the distant cities and gain perspective on who we are and what we have become."

2

The Anthropocene Challenge to Our Worldview

The insight that human activity has been transforming the metabolism of the biosphere to such an extent that it threatens the future existence of our species has provoked several kinds of reactions among different people. In the two decades since the notion of the Anthropocene was introduced, at the start of this millennium (Crutzen and Stoermer 2000), it has generated a great number of books and articles discussing its various implications. Some writers have traced the history of debates regarding the naming and dating of this new geological age, referring to the research results of the Earth System sciences. Some have predicted the global environmental changes it is likely to imply over the coming decades and centuries, given different possible scenarios with regard to human economy, technology, and demography. Many have tried to identify the root causes of anthropogenic environmental degradation, offered political proposals for mitigating it, and deliberated on why such proposals are not being adopted. Others have suggested ways in which the concept of the Anthropocene should be changing mainstream understandings of the relation between nature and society, and of what it means to be human. Whether approaching the topic from the perspectives of natural science, social science, or the humanities, a great number of people concerned about our future have increasingly found their thinking framed in terms of the far-reaching implications of the Anthropocene.

The Anthropocene obviously and unavoidably merges empirical facts, political perspectives, and existential issues. The knowledge that human activities have not just transformed but also *degraded* much of the biosphere – as evident, for instance, in soil erosion, air and water pollution, and loss of biodiversity – has been a recurrent source of remorse for millennia, but the expanding ecological impacts of human societies did not mobilize mass political protests until the emergence of the environmental movement in the 1960s. Over the final three decades of the twentieth century, rising concerns about those impacts became a prominent feature of public debate and political rhetoric. It was not until the turn of the millennium, however, that the full magnitude of our predicament became apparent.

Epitomized in the acknowledgment of anthropogenic climate change, the negative environmental impacts of human societies are no longer just an issue mobilizing a social movement, but a matter that most governments in the world profess to take very seriously. The main reason for this shift is the next to universal consensus among natural scientists that current levels of greenhouse gas emissions, if unabated, will lead to disastrous global warming. A major share of these emissions derives from the combustion of fossil fuels. This insight poses an enormous and seemingly unsolvable dilemma for the world's politicians. On the one hand, they tend to realize that the massive combustion of fossil fuels cannot continue, but on the other, they discover that the kind of societies and policies that they hope to promote – and that their voters expect – appears to be inextricably based on fossil energy. Contemporary ideals regarding globalized economic growth, free trade, agroindustry, consumerism, and automobility are unthinkable without fossil fuels, but no feasible technological alternative with which to substantially replace fossil energy appears to be available. This impasse, which explains the conspicuous incapacity of governments to deal with climate change, forces us to rethink some fundamental aspects of our modern worldview.

Whereas the profound critiques of modern civilization launched ever since the Industrial Revolution have generally been categorized – and dismissed – as simply representing a contradiction between different political positions, current concessions regarding climate change reflect a quite new situation. In contrast to earlier periods, the expertise of the natural sciences now overwhelmingly provides support for a radical critique of the dominant model of human society. The emergent worldview of the Anthropocene is reorganizing our way of drawing boundaries between the natural and the social. Never has it been so apparent that the economic models and doctrines that guide our politicians cannot be cultivated in isolation from natural science. This objection to mainstream economics has repeatedly been raised over the past two centuries (Martinez-Alier 1987), but it can no longer be ignored. Conversely, as the mirror image of this necessary acknowledgment of nature in economics, the science of engineering will have to acknowledge the role of world society in accounting for the phenomenon of modern technology (Hornborg 2001, 2013a, 2016). As they tend to imply a societal redistribution of demands on human labor time and natural space, the technologies that make so-called *energy transitions* possible need to be theorized from the perspective of social science (see Chapter 7). This is as evident when we consider the beginning of the fossil-fuel era – the Industrial Revolution – as when we consider the prospects and implications of bringing it to a close.

The underlying conviction from which I approach the confusion of the Anthropocene is that the modern worldview that so obviously has failed to deal with climate change is fundamentally flawed. It was consolidated in the core of the British Empire at the time of its historical zenith, and its two central assumptions remain foundational to most modern politics today: (1) economics is concerned with money and market prices and does not need to consider nature; and (2) engineering is concerned with inventions for harnessing natural forces and does not need to consider the structure of world society. This distinction between the "economy" as sequestered from nature and "technology" as sequestered from society is now being challenged by the Anthropocene. That is why the implications are so confusing.

One way of accounting for the persistence of such fundamental illusions is by asking how ignorance is created and maintained (Proctor and Schiebinger 2008; Bonneuil and Fressoz 2015). How are false or misguided interpretations established? This is an illuminating way of approaching the history of ideas as well as modern sociology of science and research policy. Much as contemporary denial of climate change may be cognitively and ideologically geared to market fundamentalism – because the acknowledgment of climate change would be to acknowledge that there is a fundamental flaw in the market (Oreskes and Conway 2008) – there were undoubtedly powerful incentives in nineteenth-century Britain to sequester the logic of the market and the design of steam engines from the eroding soils and human misery of colonial cotton plantations. Rather than depoliticizing environmental issues in the past – and thus obstructing the understanding of present issues – we must understand how our current predicament was created "*despite* very consistent warnings, knowledge and opposition" (Bonneuil and Fressoz 2015: 78–79; emphasis in original).

A growing number of economists are realizing that market prices do not register the biophysical processes on which the economy is based

(Spash 2017). A corollary of this insight is that monetary accounts of world trade do not reflect the asymmetric transfers of biophysical resources that commodity flows represent (Dorninger and Hornborg 2015; Hornborg and Martinez-Alier 2016). A logical implication of such ecologically unequal exchange is that the uneven accumulation of technological infrastructure – and the very existence of modern technology – is contingent on asymmetric resource flows. However, our conventional misunderstanding of technological development – as a morally neutral index of progress – is difficult to expose, undoubtedly because it requires us to view tangible material forms as manifestations of intangible societal exchange relations. Most people acknowledging the problem of climate change – regardless of whether they are climate activists championing solar power or ecomodernists championing geoengineering – thus share a faith in technological solutions. This faith is founded on the assumption that technology is nothing but revealed nature: politically and morally neutral inventions without any necessary distributive implications. To expose how flawed such an assumption is, we need to break down the conceptual barrier between nature and society in our modern worldview, and to acknowledge the extent to which they are intertwined in material phenomena such as technologies.

A model for such interdisciplinary thinking was the economist Nicholas Georgescu-Roegen (1906–1994), who is recognized as a source of inspiration for ecological economics as well as the movement for *degrowth*. A central insight for which he is remembered is that economic processes such as industrial production inexorably increase physical *entropy* (Georgescu-Roegen 1971), which is a measure of disorder or randomness in the distribution of energy and matter. This biophysical aspect of the economy remains completely ignored by mainstream economics, yet it is fundamental to grasping why the diffusion of carbon in the atmosphere is an inevitable consequence of fossil-fueled industrialism. Nor did Georgescu-Roegen view the negative environmental impacts of the economy as a merely contingent disadvantage that could be alleviated by switching to a cleaner technology: he argued that the notion that solar power can replace fossil energy is a myth, underestimating the huge volumes of matter that would be required to harness the diffuse energy of sunlight (Georgescu-Roegen 1986). It is another matter, I would add, that economically privileged nations are able to displace such physical obstacles to areas of the world where the costs of materials and labor are significantly lower. This merely illustrates the extent to which the feasibility of technological progress is contingent not just on advanced engineering but also on global asymmetries in purchasing power. Other examples include biofuels, electric cars, and similar proposals for replacing fossil energy with more sustainable technologies. The feasibility of any globalized technology is inherently dependent on relative market prices and the asymmetric resource flows that they make possible.

The Industrial Revolution inaugurated the Anthropocene by connecting the accumulated energy deposited in the past by the metabolism of the biosphere with the metabolism of human societies. For more than three billion years the metabolism of the biosphere has been based on photosynthesis. Beginning a mere 250 years ago, by means of the steam engine and later the combustion engine, human societies began accessing the energy accumulated over hundreds of millions of years in geological sediments. These technologies, by converting accumulated solar energy into self-reinforcing societal processes, are what has made the connection between photosynthesis and the Anthropocene possible. Whereas the control of fire has long given humans access to such energy for heating, the steam engine represents the first use of products of photosynthesis – whether fossil fuels or firewood – for *mechanical* energy, that is, as substitutes for muscular work. From bands of hunter-gatherers to empires based on slavery, the organization of labor is a central concern of human societies; hence, the large-scale use of labor-saving machinery is obviously not merely a technical but also a thoroughly *social* phenomenon. In combination with the operation of an increasingly globalized market, it has historically redistributed workloads among different social groups and sectors of world society. These fundamental societal transformations simultaneously had momentous ecological implications, displacing environmental loads among regions and in some areas relieving the surface of the land from the imperatives of yielding energy (Wrigley 1988; Sieferle 2001). Although the notion of "technological progress" since the Industrial Revolution essentially reflects the capacity of human societies to harness fossilized sunlight, this capacity has for two and a half centuries been inextricably linked to global asymmetries in purchasing power, and ultimately also implies the capacity to appropriate, by means of the market, the products of photosynthesis from land surfaces in other parts of the world. The use of inorganic energy technologies, in other words, provides access to organic energy such as biomass and human labor elsewhere. The Anthropocene thus prompts us to rethink the modern concept of technology, which tends to raise expectations of a magical human ingenuity that promises to liberate us from all kinds of limitations, but that involves a less obvious, distributive dimension that is essentially social.

Current notions of human progress tend to hinge on economic growth and technological innovation. This outlook seems to lead some people to think that the issue of whether the planet will be inhabitable a hundred years from now is subordinate to indications that an increasing share of the world's population is modestly improving its health, education, and purchasing power. In this view, in other words, it does not seem to matter so much if we are generating changes that will lead to the extinction of our species, if increasing numbers of people today live somewhat longer, spend more years in school, and are able to consume a bit more than their parents (cf. Norberg 2016; Pinker 2018; Rosling et al. 2018). The illusion projected by the proponents of such a definition of progress is that there is a contradiction between sustainability and equality. But this is paradoxical because the current model of global development is *both* unsustainable and increasingly unequal. The gap between the wealthiest and poorest segments of world society continues to widen (Alvaredo et al. 2017). Genuine progress would be to recognize that the ongoing displacements of work and environmental loads, by means of markets and technology, to poorer parts of the world are no less defensible than slavery. However, to halt their insidious logic requires not just reconceptualizing the ontology of markets and technology but also nothing short of problematizing – and redesigning – the very artifact of money.

Money and the Anthropocene Predicament: Reflections on Some Views from the Human Sciences

The burgeoning literature on the Anthropocene can be divided into three main genres: (1) Earth System science, generally focused on analyses and predictions of global climate change based on natural-science research; (2) social-science analyses of various societal implications of the findings of Earth System science, often combined with proposals for political action to mitigate climate change; and (3) reflections on the philosophical and existential implications of Earth System science and the notion of the Anthropocene. In what follows I shall not attempt to summarize the findings and predictions of the natural scientists, but will briefly discuss a few contributions to the latter two genres.

The increasingly alarming degradation of the biosphere was apparent decades before the term *Anthropocene* was proposed. In 1992, more than 1,700 scientists signed a text called "World Scientists' Warning to Humanity," reminding readers that "human activities" were inflicting "harsh and often irreversible damage" on the atmosphere, water resources, oceans, soil, forests, and species diversity (Union of Concerned Scientists 1992). Twenty-five years later, in December 2017, more than 15,000 scientists from 184 countries signed a renewed warning, showing that, by and large, these alarming trends had been exacerbated over the past quarter-century, particularly emphasizing the "current trajectory of potentially catastrophic climate change" due to greenhouse-gas emissions from burning fossil fuels, deforestation, and agricultural production (Ripple et al. 2017). One of the signatories of the 1992 warning was Paul Crutzen, who in 2000 co-authored the influential proposal that the current geological epoch should be termed the Anthropocene (Crutzen and Stoermer 2000).

Over the past five decades, enormous amounts of research, debate, and political negotiations have been devoted to the human-driven deterioration of the biosphere, but the destructive processes tend to continue. Out of the nine significant "planetary boundaries" that need to be respected not to jeopardize human well-being, four are currently being transgressed, namely climate change, rates of biodiversity loss, land conversion, and the loading of nitrogen and phosphorus (Rockström et al. 2009; WWF 2016). Although from its invention the concept of the Anthropocene has been linked to concerns over global sustainability, it has generated a voluminous academic debate, often focusing on the legibility of its onset in the geological record, about when in human history the postulated new epoch began. There have been proposals for dating the Anthropocene to the Pleistocene extinctions of megafauna beginning about 50,000 years ago, the origins of domestication around 10,000 BC, the so-called Columbian exchange and depopulation of the Americas following 1492, a patent on steam engine design from 1784, and the first nuclear detonation in 1945, which simultaneously marked the beginning of the "Great Acceleration" of the global economy (Bonneuil and Fressoz 2015; Angus 2016; Davies 2016; Hamilton 2017).

In view of the possibly catastrophic implications of currently ongoing anthropogenic environmental change, the most significant challenge can hardly be to robustly date the origins of such change – in accordance with the principles of geology – or even to give the new geological epoch an appropriate name, but to understand what is propelling the forces of destruction and hopefully also how they might be mitigated. The incapacity of modern societies to control these forces indicates either that such an understanding has continued to elude them, or that powerful interests are – perhaps inadvertently – obstructing potentially successful political strategies to do so. It is reasonable to conclude that our impasse derives from a combination of these two explanations. There is a subtle and undoubtedly subconscious symbiosis between power and knowledge that explains why the articulation of new perspectives can be a truly revolutionary activity.

There has been no shortage of eloquent accounts of the ethical, political, and ontological challenges of the Anthropocene. Much of the literature emphasizes the imperative of human responsibility and restraint, without specifying whether these troubled pleas are aimed at politicians or people in general (Davies 2016; Hamilton 2017). Some of it unequivocally and indignantly defines the root problem as an abstract economic system referred to as capitalism, which can only be overthrown through the mobilization of mass movements against a class of people in whose interests it is to reproduce the system (Foster, Clark, and York 2010; Klein 2014; Angus 2016; Malm 2016). Other writers have been content to advocate a radical – and frequently very well-argued – rethinking of economics, without positing the necessity of revolution (Victor 2008; Jackson 2009; Raworth 2017). Yet others dwell on various ontological implications of our transformed understanding of human-environmental relations (Haraway 2016; Danowski and Viveiros de Castro 2017; Latour 2017). Whatever their particular preoccupations, it has struck me that the many commentators on the human dimensions of our global predicament have one thing in common: they devote no serious attention to the most significant and transformative offspring of our species' capacity for abstract representation, namely *money*. Yet the idea of money is arguably the ultimate root of the Anthropocene predicament.

Although unique to our species and the current historical period, general purpose money is almost universally taken for granted, probably because it has become "second nature" to the extent that it is not seen as something about which we even have a choice. Given its ubiquitous role in our everyday lives, it increasingly appears fruitless to assign responsibility for its aggregate implications to any particular social group, such as politicians, economists, or capitalists. The logic of its aggregate implications has been abundantly documented and criticized for centuries, but it has so far proven unproductive to blame its negative repercussions on the discipline and profession whose task it has been to theorize that logic, or even on a class defined as its primary beneficiaries. The situation is indeed analogous to a game of *Monopoly*, in which frustrations are directed at winning players, or at those particularly well acquainted with the rules, while no one fundamentally questions the rules. The Marxist tradition has certainly challenged the rules of the game of capitalism – and

provided some very incisive accounts of its insidious ecological consequences (e.g., Foster et al. 2010; Klein 2014) – but possibly without fully
realizing the extent to which money is the essential motor of that game
(Moseley 2016). As illustrated by several political experiments over the
course of the twentieth century, a postcapitalist society using general
purpose money is a contradiction in terms. Many nation-states have
attempted to curb the deleterious logic of money within their territories
but discovered that globalization – simultaneously an indispensable
source of economic growth and a constraint on their political agency –
is the capacity of that logic to transcend territory, encouraging maximum
"efficiency" defined in terms of promoting the lowest wages and the least
regard for the environment. Widening economic inequalities and accelerating environmental degradation can thus be derived from the increasingly global reach attained by the logic of money. Meanwhile, it is ironic
that the social theorists who ought to be best equipped to trace the
insidious logic of capitalism to the "agency" of a specific artifact – the
proponents of Actor-Network Theory – have chosen to turn their attention elsewhere (Latour 2017). Nonetheless, as I elaborate in Chapter 13,
the ultimate challenge of the Anthropocene is to envisage a social system
that is not organized in terms of the inertia inherent in the artifact of
general purpose money.

Clive Hamilton has scrutinized the evidence from Earth System science
and contemporary politics, finally admitting to himself that "we simply
are not going to act with anything like the urgency required," which
relieved him of some of the anger he had felt at "the politicians, business
executives and climate sceptics who are largely responsible for delaying
action against global warming until it became too late" (Hamilton 2010:
xiii). He traces the disastrous inertia of modern society to "growth
fetishism," consumerism, psychological denial, alienation from nature,
and trust in improbable technological solutions. Genuine despair and
acceptance of catastrophe may be requisite to the "revived democracy"
that he still hopes for: "the mobilisation of a mass movement to build a
countervailing power to the elites and corporations that have captured
government" (ibid., 218). In the midst of his despair, Hamilton thus
sympathizes with the goal of "learning how to live wisely, cooperatively,
and well within the limits of the Earth, while at the same time cultivating
our potentialities and those of the planet" (Hamilton 2017: 114). The
facts of the Anthropocene prompt us to adopt a new kind of anthropocentrism, he argues, not the arrogant kind expressed in ecomodernist
dreams of geoengineering, but rather a "humble anthropocentrism" that

assumes the burden of responsibility that adheres to power (ibid., 43–44). The democratic exercise of collective self-restraint, he suggests, might help us regain a sense of meaning to history and to human life (ibid., 116–120). Against the background of my previous arguments, I can see no more crucial and meaningful way of exercising such collective self-restraint than to consciously redesign the artifact of money, the logic of which is currently driving what Hamilton calls "the hydra-headed monster known as globalization" (ibid., 82), aggravating climate change, global inequalities, and vulnerabilities of various kinds.

In their meticulously documented book *The Shock of the Anthropocene*, Christophe Bonneuil and Jean-Baptiste Fressoz (2015: 70) highlight the roles of social actors and political strategies that over the past two centuries have generated – and narrated – global environmental change. They demonstrate how dominant narratives have supported various policies that have intensified resource use and environmental degradation. Whether celebrating rising energy use, consumerism, war, free trade, colonialism, or mechanization, narratives such as neoclassical economics and the modern ideology of progress have systematically disarmed and marginalized resistance to environmental destruction, industrialization, alienation, and imperialism, both in Euro-American core nations and their dependent peripheries. Bonneuil and Fressoz emphasize that the silencing of such recurrent and often desperate environmental warnings and critique over the past centuries has been conducive to the current narrative of the Anthropocene, which suggests that the ongoing transgression of environmental limits is a recent discovery, after centuries of "innocent" Euro-American expansion. This is an important reminder, honoring the protests of myriad distressed voices long silent, but it does not explain why so many established scientific specialists are now suddenly confronting the politicians and corporations head-on. As Hamilton observes, the trajectories of climate change that are now being traced by Earth System scientists evoke apocalyptic environmental destruction at a planetary scale, which was unimaginable a few decades ago (Hamilton 2017: 19–20). Although local or regional degradation has frequently threatened the survival of communities and ethnic groups, drastic disruptions of the Earth System imply a threat to the entire human species.

Ian Angus (2016) has provided nonspecialists with a particularly lucid overview of the current state of Earth System research on the implications of different emission scenarios for climate change. He clearly demonstrates the pivotal role of fossil energy in the alarming trajectories of the Anthropocene. However, as much as I would like to believe in his vision of a political alternative to apocalypse, I suspect that he confuses the symptoms of unsustainability with their source. He begins part three of his book *Facing the Anthropocene*, called "The Alternative," by quoting Gus Speth's disillusioned conclusion, after 40 years of working inside the system, that "the capitalist system itself is the cause of environmental destruction" (ibid., 189). The "features of capitalism" that Speth denounces include the commitment to economic growth, investment in environmentally harmful technologies, the corporate objective to generate profit, markets, government committed to growth and subservient to corporate interests, and rampant consumerism spurred by advertising. But my question is if "capitalism" is really the ultimate agent here?

My reason for posing this question is that *all* the deleterious features and incentives listed by Speth can be traced to the logic encouraged by general purpose money. No nation on Earth is unafflicted by them. Is it really fruitful to blame them on an abstract "system," rather than derive the destructive tendencies from the global common denominator that shapes virtually everyone's behavior? Although I completely agree that our goal must be to build an "ecological civilization" (ibid., 196), I believe that the political history of the past two centuries confirms that it is unrealistic to build a society on principles that systematically counteract the incentives inscribed in the very artifact that organizes the everyday lives of its citizens. What kinds of schizophrenic legislation would such a society require, and what kinds of administrators? An ecological civilization can only be built by circumscribing the globalized horizons of commensurability that are inherently codified in conventional money. To be able to democratically realize this goal, we must begin by agreeing that much of the purchasing power granted to Euro-Americans in remuneration for their labor must be constrained from being converted into the products of much lower-paid labor and environmental destruction on the other side of the planet – and the transports of these products across the globe. This is *not* a less realistic goal than overthrowing the "capitalist system." As I see it, capitalism *is* the aggregate logic of general purpose money. To permit the logic of generalized commensurability – inscribed in the conventional artifact of money – to run its relentlessly globalizing course is to encourage continuously aggravated global inequalities, climate change, and various other forms of vulnerability. It is not difficult to imagine that this might indeed lead to the end of the world. But then again, there is always the chance – particularly when things get worse – that sufficient numbers of people will see the potential benefits of changing the rules of the game that we are currently playing.

Fundamental to those rules are the properties of the very tokens that we use as everyday means of payment. It is not at all inconceivable, at least in principle, to design them so as not to allow us to contribute to the impoverishment and ecological degradation of populations far beyond our horizons.

Considering Some Propositions from Actor-Network Theory

This intervention might be taken to indicate that, like the proponents of Actor-Network Theory, I am attributing "agency" to artifacts. This is in fact *not* my position. But I believe that it may be as misguided and fetishistic to attribute agency to abstract systems as to inert objects. The features we inscribe in artifacts tend to have systemic consequences at the aggregate level of society, but to transform their systemic logic we need to modify those features. To consider the relations among humans, their artifacts, and their socioecological systems, I shall discuss the perspectives of Bruno Latour – a central proponent of Actor-Network Theory (Latour 2005) – on the ontological implications of the Anthropocene. In contrast to the analytically clear and straightforward prose of the Earth System scientists and most of their human-science exponents (e.g., Bonneuil and Fressoz 2015; Angus 2016; McNeill and Engelke 2016; Hamilton 2017), Latour's style of writing can be tortuously verbose and imprecise, which makes it difficult to identify concrete arguments with which to engage. Nevertheless, as he is immensely influential in the human sciences, and particularly in the so-called environmental humanities – promotions of his recent books *Facing Gaia* and *Down To Earth* call him "one of the world's leading sociologists and anthropologists" (Latour 2017, 2018) – I shall try to distil and comment on some of the points I believe he is making.

In *Facing Gaia*, Latour applies many of the perspectives on the human condition for which he has become famous. I shall here focus on three of his claims over the years, which have been particularly provocative: (1) scientific and other understandings of natural phenomena are constructions or fabrications shaped by the mutual interaction between observers and the observed; (2) living and nonliving components of a network all have comparable capacities for agency; and (3) distinctions between society and nature and between subject and object are mere constructions of modernity that obscure the mutual entanglement of multiple actants. All these three claims, which he has defended in highly publicized debates such as the so-called Science Wars, become no less problematic when applied to the dilemmas of the Anthropocene.

With regard to the first claim, Latour admits that it has been embarrassing to find the "climate sceptics" mobilizing similar, constructivist arguments to undermine the findings of Earth System science. He appears to be thoroughly persuaded about the reality of climate change. As Hamilton has remarked, "[T]he Anthropocene brings us back [from constructivism] to Earth with a thud" (Hamilton 2017: 79). Yet it is difficult to follow Latour's evasive argument suggesting that his assertion that scientific facts are fabrications does not mean that they are not true. If the findings of Earth System science are indeed true, would it not imply that there is an objective reality out there that can be more or less successfully represented? How would Latour have wanted to modify the research methods of the Earth System scientists? What, in retrospect, was the point of the position he struggled to defend during the Science Wars? He continues to want to abandon a distinction between objective facts, on the one hand, and normative prescriptions, on the other, repeating that it would be incorrect to believe that the latter could be derived from the former, but does this mean that he refutes the significance of impartial scientific research as a foundation for prescriptions such as abandoning fossil fuels, chlorofluorocarbons, or cigarette smoking? Do such prescriptions imply partiality? Is the efficacy of scientific methodology contingent on the *concerns* of the scientists?

With regard to the second claim, the onset of the Anthropocene has prompted Latour to gravitate toward James Lovelock's famous deliberations on *Gaia* (Lovelock 2000). The lecture series on which Latour's book is based is largely devoted to contemplating in what sense planet Earth can be considered a locus of agency. These deliberations draw on his previous arguments on the purported agency of peptides, meandering rivers, and other entities not conventionally classified as agents. As I show in Chapter 10, however, there is a huge difference between the functional components of living organisms, on the one hand, and rivers, on the other. The homeostatic agency inscribed in every aspect of a living organism, honed by millions of years of natural selection, is something entirely different from the purposeless trajectories of a meandering river or a heating planet. Latour suggests that the changes in the Earth System provoked by emissions of greenhouse gases are inverting the traditional relation between the human subject and the natural object, so that what earlier used to be thought of as an object (planet Earth) has now emerged as a subject (Gaia) (Latour 2017: 73). His use of the concepts "subject" and "object" appears to be exclusively political, in the sense of designating a hierarchical relation between an active and a passive part, rather than *ontological*, in the sense of indicating whether an entity does or does not have a capacity for sentience, as is one meaning of the cognate concepts "subjective" and "objective." My Oxford dictionary provides definitions of *subject* and *object* that confirm the latter usage, namely *subject* as "a thinking or feeling entity" and *object* as "a thing external to the thinking mind or subject." This distinction appears to be quite irrelevant to Latour, who asserts that "signification" is a property of all the phenomena he classifies as agents, whether army generals, rivers, peptides, or gravity (ibid., 69), completely oblivious of Eduardo Kohn's (2013: 92) explicit admonition to Actor-Network Theory that only sentient "selves" qualify as agents. The category of "nonhumans" – frequently invoked by Latour and other posthumanists – is a crude concept, encompassing a strange and heterogeneous mix of "microorganisms, animals, plants, machines, rivers, glaciers, oceans, chemical elements, and compounds" (Danowski and Viveiros de Castro 2017: 100). The exploration of Amazonian animism by anthropologists such as Viveiros de Castro and Kohn has highlighted an inclination among nonmodern people to acknowledge the subjectivity of nonhuman animals, but the tendency among posthumanist academics to impute to nonmoderns a complete dissolution of the difference between subjects and objects is a distortion of the ethnographic evidence (Viveiros de Castro 1998; Santos-Granero 2009; Kohn 2013: 7, 91).

With regard to the third claim, the "Nature/Culture schema" that Latour attributes to the modern worldview is indeed problematic. I do not believe that anyone today subscribes to the notion that nature represents a sector of the world that is "objective and inert," while culture is "subjective, conscious, and free" (Latour 2017: 85). If instead taken to denote all *aspects* of reality that do not derive from human culture (or society), that is, that do not require explanations referring to the symbolic capacities of human beings, nature obviously generates myriad entities that are animate and subjective, while culture (or society) continuously produces countless inert objects. Latour seems to assume that the nature/culture distinction and the object/subject distinction coincide, but this is quite unwarranted. Exclusively natural processes account for billions of sentient creatures as well as geological formations, while cultural or social factors yield inanimate artifacts as well as human consciousness. With the global expansion of human economies, purposes, and designs, the adjectives "natural" versus "social" increasingly refer to analytical *aspects* of phenomena, rather than to concrete domains that can be physically delineated. This does not mean that "the notion of 'nature' [is] as obsolete as that of 'wilderness'" (Latour 2017: 121) but that it needs to be reconceptualized in these terms. Latour has demonstrated that our inclination to purify socioecological phenomena as belonging exclusively to nature or society is misguided, but this valid observation does not justify a dissolution of the concept of "nature." As some of us have been trying to remind environmental philosophers over the years, nature has not vanished just because it has been mixed with society (Hornborg 2015; Maris 2015; Crist 2016; Malm 2018).

It was incorrect of Lovelock (2000: 56) to assert that "the Gaia system shares with all living organisms the capacity for homeostasis" because the Earth System has not been set to regulate itself by a process of natural selection. Latour (2017: 95) asks what difference there is "between the status of 'sentient being' and that of 'self-regulated system'," but this should be obvious to anyone who can distinguish between a living organism and a market or cybernetic machine. Latour is wrong in asserting that every organism "intentionally" (ibid., 98–99) modifies its environment to increase the likelihood of its own survival; although its environmental impacts are sometimes beneficial to an organism, there is no intentionality or teleology at work. Latour's systematic efforts to dissolve the distinction between the purposeful agency of an organism and the complex reactivity of the Earth System become particularly transparent when he imputes "sensitivity" to Gaia (ibid., 141). Reactivity is not synonymous to sentience. Similarly misguided are his apparently serious deliberations on how to grant phenomena such as "Oceans," "Lands," "Atmosphere," and "Forests" representation and voice in a democratic parliament of *things* (ibid., 255–270). Not only is his enthusiasm about representational democracy surprising in a world where, for instance, the (human) voices of "Indigenous Peoples" are hardly comparable to that of "Cities" or "Economic Powers" but also the obvious impossibility of granting voice to nonhuman components of the biosphere makes the whole exercise – although logically consistent with Actor-Network Theory – as ridiculous and politically naïve as the theory that inspired it.

In sum, Latour's deliberations on the agency of Gaia in the Anthropocene are far from persuasive. Nevertheless, his Actor-Network Theory provides us with two important insights that deserve a more succinct and politically coherent articulation: (1) the design of our artifacts can impose a specific systemic logic on the socioecological contexts in which we participate; and (2) our historical inclination to purify socioecological phenomena as deriving exclusively from either social or natural processes fundamentally distorts our understanding. It is unfortunate and paradoxical that Latour's wide-ranging discussions on the distribution of agency – which he identifies as "basically the only subject" (ibid., 271) of his lectures on Gaia – never touches upon the capacity of the money artifact to transform the functioning of the Earth System. Equally disappointing is his failure to theorize the systematic incapacity of economics and engineering science to deal with climate change as paradigmatic illustrations of our conventional purification of the economy as sequestered from nature, and of technology as sequestered from society.

There is undoubtedly some cosmological affinity between the market's denial of nature and the "ontological dissolution" of nature by constructivism (Bonneuil and Fressoz 2015: 220). The anguish and shock of the Anthropocene for constructivist human science is the discovery that the world is as real and vulnerable as our mortal bodies, and that global disaster is not just an image (Danowski and Viveiros de Castro 2017). But the ultimate contradiction – and constraint – of Latour's Actor-Network Theory is its mobilization of voluminous and convoluted theoretical jargon in the promotion of politically impotent empiricism. It finally does not contribute an emancipatory theory for explaining and transcending the terrors of the Anthropocene.

The Anthropocene predicament will ultimately force us to acknowledge – contrary to a conventional modern worldview – the *natural*, biophysical aspects of what we know as economic growth, as well as the *social*, distributive aspects of what we know as technological progress. Such a shift of perspective requires that we recognize the aggregate capacity of monetary flows to physically reorganize the Earth System, not least through the accumulation of fossil-fuel technologies. The accumulation of fossil-fueled technological infrastructure in wealthier sectors of world society contributes to increasingly severe global inequalities and to rapid degradation of the biosphere. These "socionatural," "socioecological," or "sociometabolic" processes defy conventional analysis founded on the ontological sequestration of economic and technological phenomena. They thus remain beyond the pale not only of mainstream politicians but also of the movements that challenge them to invest in energy transitions built on new technologies. Most fundamentally, we shall have to acknowledge that the quintessential artifact of modernity – the meme of general purpose money – embodies an intrinsic logic that inexorably generates widening global gaps and environmental destruction. This logic is inherent in conventional money. It guides the strategies devised by economic advisors everywhere, and it frustrates political attempts to contain it. To be able to imagine the end of capitalism, we must begin to imagine ways of redesigning money. As I propose in the final chapter, the essential principle for such reforms must be to limit the spatial reach of economic commensurability by restricting a substantial proportion of circulating exchange values to purchases within a geographically defined radius. A democratic agreement to thus restrain the logic of money might curb the polarizing, asymmetric resource flows currently orchestrated by the world market, and the escalating, suicidal destruction of the Earth System.

3

Producing and Obscuring Global Injustices

Writing toward the end of World War II, the economic historian Karl Polanyi in *The Great Transformation* ([1944] 1957) showed that the "idea of the self-regulating market" for a century and a half had generated a recurrent tension between the notion of free trade and the problems of transforming people, land, and money into commodities. If society did not protect itself from its logic, he argued, the market would inexorably increase economic inequalities, environmental degradation, and financial instability. In Polanyi's view, the rampant nationalism of the two world wars, and the intervening economic crisis, were reactions to the nineteenth-century expansion of world trade.

We today have every reason to reconsider Polanyi's warning. For two and a half decades after the war, unprecedented economic growth in North America and Europe was able to accommodate various societal policies for domesticating the market. In the affluent Global North, the most damaging consequences of general commodification seemed possible to alleviate. This was the golden age of the welfare state. Critics arguing that the impoverishment of labor and land had simply been displaced to an "underdeveloped" South could be contradicted by blaming underdevelopment on its politicians. But when the financial crises of the 1970s finally prompted a new wave of liberal fundamentalism, the insidious logic of the untamed market soon became obvious. Precisely as Polanyi could have predicted, increasing inequalities, environmental degradation, and financial instability have plagued us since then. In recent years, there has even been an ominous resurgence of nationalism.

All this time, however, those of us who are passionate about making the planet a better place have held on to visions of a future world society

promoting justice, sustainability, and resilience. Whether liberal or socialist, we have imagined a world where seven billion or more people share
the kind of material security enjoyed by Scandinavians in the 1960s.
Those visions have been so important to the moral identity of large
portions of the middle class that it widely and viciously condemned the
currently widespread rejections of globalism in Britain and the United
States. But the worst moral conundrum for these middle-class segments of
the North is undoubtedly that their own lifestyle and material security is
contingent on the very polarizing and ecologically deleterious global
processes to which they are ideologically opposed. Given the widening
inequalities, soaring greenhouse gas emissions, and exponentially increasing debts, the faith in globalization and global development has lost its
credibility.

We have somehow been able to disregard the recurrent reminder that,
if European or American affluence were universalized, it would require
four additional planets. Most people still want to believe that the world is
about to abandon fossil energy, although it continues to account for
around 86 percent of global energy use and shows little sign of abating.
In similar ways, most of us seem to be able to neglect reports that our
environmental impacts have already transgressed four out of nine planetary boundaries, and that, rather than "decoupling," our economies are
continuously growing *increasingly* wasteful of materials. In 30 years'
time, if the current trend continues, the world's population will need three
times as much resources as today (Schandl et al. 2016). Moreover, within
the next 70 years, the population of Africa is expected to quadruple –
from one to more than four billion people – while climate change may
make much of the continent uninhabitable. These prospects unquestionably underscore current concerns about sustainability.

Paradoxically, the ideological bankruptcy of globalization has left even
socialists with little to hope for. Although intent on challenging inequalities and financial speculation, there is little they can do about the physical
constraints of our planet. The demands of industrial production, urbanization, international trade, and decent living standards for around 10 billion people do not suggest a bright socialist future. The sympathetic and
democratic values that have propelled the labor movement, the environmental movement, and the movement for human rights are simply not
compatible with the logic of conventional money and world trade.
Against this background, it sounds increasingly contradictory when
members of the affluent middle class in the Global North claim to want
to promote global justice and sustainability, as the very economic logic

that makes their affluence possible inexorably increases global gaps and environmental degradation. Even those of us who are most intent to save the planet find ourselves belonging to its heaviest burden but must – like elites through history – struggle psychologically to maintain our moral identity. It is among the affluent middle classes of the Global North that we find the strongest appeals to sustainability and global human rights. The most vociferous proponents of renewable energy and global solidarity thus tend in practice to have the largest carbon footprints and to be the greatest consumers of embodied low-wage labor and natural resources appropriated from the Global South.[1] The dilemma of upholding a virtuous identity has become increasingly evident as globalization has aggravated economic inequalities within nations as well as in the world as a whole, prompting less affluent segments of core nations to resent the flows of migrants from economically marginalized countries – and the middle classes to denounce these "populists" as xenophobic.

An insidious aspect of recent ideological transformations is that *any* critique of globalization risks being classified – and rejected – as "populism." In its eagerness to maintain a distance to right-wing nationalism, leftism has unwittingly been co-opted into the neoliberal camp to which nationalism is a reaction. Leftist appeals to internationalism and global solidarity have thus imperceptibly fused with the neoliberal rhetoric on globalization and the dismantling of borders. Socialists and liberals tend to be united in a middle-class, cosmopolitan outlook referred to by populists as "globalism." To criticize the globalization of the labor market is to risk being classified as xenophobic (cf. Daly 2018: 102). Paradoxically, the recent surge of nationalism in the United States and Europe may thus serve to strengthen the mainstream hegemony of neoliberal ideology. "Populist" sentiments regarding the significance of spatially defined cultural identities are predictable consequences of globalization, but the moral denunciation of such sentiments by socialists and liberals alike reflect an insufficient grasp of the existential repercussions – particularly among the economically insecure – of a dismantling of national borders (cf. Bauman 1998; Goodhart 2017). It is no doubt a major dilemma for socialists to find substantial parts of the working class not conforming with their expectations of what is to be regarded as a

---

[1] By and large, the most prominent producers of modern ideology – politicians, journalists, academics, educators, artists – tend to enjoy purchasing power and living standards that are significantly higher than the national average. They no doubt also tend more frequently to conduct international flights.

decent class struggle. Their understanding of the dynamics of social change has obviously underestimated the phenomenon of cultural identity.[2] The disruptive effects of globalization on identity formation – and their transformation into nationalism – is yet another problematic consequence of general purpose money.

If there is to be a third option alongside the choice between nationalism and globalization, it would mean recognizing that the operation of markets and money is no less socially constructed than the mobility of a chess piece, but that such constructions have very tangible social and material consequences. To acknowledge the extent to which the destiny of human society and the biosphere has been delegated to the mindless cybernetics of our fetishized artifacts is like snapping out of a delusion. To fathom the implications of this delusion would make us more receptive to the idea of a radically reorganized economy. But it requires that we fundamentally rethink what an economy is, so that we are prepared to see our conventional economic categories as obstacles to perceiving what is happening when people exchange things with each other. The way in which human societies organize exchange is fundamental to the conditions for economic inequalities as well as for identity formation. To rethink what an economy is, we must begin by reconsidering the material implications of market exchange.

Ecologically Unequal Exchange

The concept of unequal exchange has been used by theorists of imperialism and world-systems to denote a transfer of "surplus value" from less to more affluent economies. In an argument couched in the Marxist labor theory of value, Arghiri Emmanuel (1972) demonstrated that differences in the price of labor (i.e., wages) between nations will generate net transfers of embodied labor in international trade. Immanuel Wallerstein (1974–1989) is generally less explicit in his use of the concept of unequal

---

[2] In recent decades there has been a noteworthy shift in the discourse on place and identity, ultimately geared to globalization. Local or place-based identities, and processes of cultural identity formation in general, are no longer approached as anthropological observations on recurrent cultural phenomena (cf. Barth 1998) but as morally reprehensible attitudes, at least if expressed by Europeans. This shift prompts us to ask if the critiques of modernity voiced by indigenous peoples, peasants, bioregionalists, and environmental justice activists are suddenly rendered obsolete. Are the virtues of small-scale subsistence practices, traditional ecological knowledge, and the cultivation of community and a "sense of place" now to be dismissed as romantic delusions?

exchange, but also draws on Marxist theory and appears to denote a transfer of surplus value defined in terms of invested labor time.

The sociologist Stephen Bunker (1985) pioneered a definition of unequal exchange that transcends the labor theory of value. His approach to development theory is highly inspired by structuralism, dependency, and world-system perspectives, but is critical of their emphasis on labor value. Using the history of the Amazon basin as a case study, he instead suggests that the polarization of core versus peripheral sectors is the result of an unequal exchange of "energy values" extracted from nature and transferred to centers of accumulation. As his contribution is pivotal for the subsequent conceptualization of *ecologically* unequal exchange, we shall consider it in some detail.

Bunker asserts that "the Amazon's major export economies have been based on extraction of value from nature rather than on the creation of value by labor," and that "the differences between extractive and productive economies [are] more fully accounted for by the laws of thermodynamics than by theories of politically enforced unequal exchange" (ibid. 12). He states that "the exploitation of natural resources uses and destroys values in energy and material which cannot be calculated in terms of labor or capital" (ibid. 22). Importantly, he notes that theories of imperialism, world-systems, dependency, and "unequal exchange based on wage or productivity differentials ... have all acknowledged primary material export as a defining characteristic of most forms of underdevelopment, but they have not systematically explored the internal dynamics of extractive systems" as a consequence of energy flows (ibid. 30). Because dependency and world-system perspectives "draw too much on labor theories of value," he asserts, "they cannot account for the multiple effects of unbalanced energy flows between different regions.... Any theory of international exchange which measures commodity flows between regions only in terms of capital, prices, or the labor incorporated into each is therefore ... wrong" (ibid. 245–246).

Bunker's interpretation of the problem is that "nature, values in nature, and the economies which depend on values in nature have been systematically undervalued" (ibid. 31). He proposes that if we widen Emmanuel's concept of unequal exchange, "then we can say that countries where labor values and natural values are seriously undercompensated will tend indeed to be underdeveloped. ... Ultimately," he concludes, "the need is to slow the flow of energy to the world center, [primarily] by raising the relative monetary costs of extracted commodities and thus slowing their consumption in the core" (ibid. 252–253).

When I published my argument on "the thermodynamics of imperialism" (Hornborg 1992) I had not yet discovered Bunker's work, but I similarly believed that value should be redefined in terms of energy.[3] However, when I later proposed an "ecological theory of unequal exchange," I emphasized that energy and values should not be confused, explicitly disagreeing with this aspect of Bunker's analysis (Hornborg 1998a). The issue is pivotal to rigorously theorizing ecologically unequal exchange. Like the fundamental Marxian notion that monetary exchange values conceal asymmetric flows of "use values," we tend to assume that there is an objective, insufficiently compensated measure of economic value, rather than simply observing that flows of money represent the asymmetric exchanges of biophysical resources on the world market *as if they were reciprocal*. The misleading argument on asymmetric transfers of underpaid "values" recurs not only in Marxian attempts to conceptualize ecologically unequal exchange and the appropriation of nature (Foster and Holleman 2014) but also in the work of some ecological economists concluding that asymmetric exchange should be expressed in terms of insufficient monetary compensation (Odum and Arding 1991; Odum 1996; Martinez-Alier 2002).[4]

The analytical convergence of labor and energy theories of value is not coincidental (Lonergan 1988). Their common source appears to be our adherence, ever since Marx, to an essentialist conception of value, that is the belief that exchange values have (or should have) an analytically specifiable relation to some objective, material substratum of reality. Paradoxically, however, a coherent theory of ecologically unequal exchange must simultaneously emphasize the economic relevance of biophysical parameters for human economies *and* disconnect them from notions of economic value. The significance of asymmetric transfers of material resources is not that they represent underpaid values, but that they contribute to the physical expansion of productive infrastructure at the receiving end. The accumulation of such technological infrastructure may yield an expanding *output* of economic value, but this is not equivalent to saying that the resources that are embodied in infrastructure have an objective value that exceeds their price. The unequal or asymmetric

---

[3] Or rather *exergy* (the energy available for harnessing for work). The same analytical flaw recurs in Hornborg 1998b, reprinted in Hornborg 2001: 57–59.

[4] In fairness to Martinez-Alier, however, it should be added that he clearly recognizes that the incommensurability of physical deterioration and monetary values "makes it hard to produce a measure of ecologically unequal exchange, in the fashion that conventional economics is familiar with" (Martinez-Alier 2002: 216).

exchanges on the world market that need to be acknowledged are exchanges of biophysical resources, not values. If our conclusion would be that the problem is that "natural values" are underpaid or even unpaid (cf. Moore 2015), our analysis would not probe deeper than the claims of mainstream economists that environmental externalities (such as "ecosystem services") are insufficiently reflected in market prices. In short, we need to liberate our analyses from the conceptual constraints of money. Monetary value is a cultural veil that obscures material asymmetries by representing unequal exchange as reciprocal.

This is not to say that it is illegitimate of victims of environmental injustice to demand monetary indemnification, as exemplified by the notion of an "ecological debt" owed by wealthier nations to less developed ones (Martinez-Alier 2002: 213–251), only to remind ourselves that money cannot neutralize ecological damage in a physical sense. Monetary compensation for environmental damage can reduce contemporary grievances, but it is illusory to believe that "correct" reparations could be calculated, or that they would somehow set things straight. The ecological debt of Britain, for instance, is as incalculable as its debt to the descendants of West African slaves. To raise the price of energy and raw materials, as Bunker suggested, would undoubtedly reduce the current magnitude of ecologically unequal trade (and the accumulation of technological infrastructure) in the world, but it would not make trade equal. Like the notion of making "correct" recompense for past asymmetries, pricing resources high enough to neutralize the damage caused by their extraction would be tantamount to shutting down industrial capitalism.

Environmental Justice

The argument so far has been that the accumulation of technological infrastructure, or what is widely referred to as development and progress, is an inherently moral and political issue. The uneven global distribution of such infrastructure is visible on satellite images of nighttime lights and closely correlated with the geographical distribution of high average GDP, that is, the distribution of money (Figure 3). The pattern agrees well with our proposal that the phenomenon of modern technology should be understood not simply as an index of ingenuity, but as a social strategy of *appropriation* (of labor and land), which is tantamount to saying that it is a strategy of *displacement* (of work and environmental loads). This capacity for appropriation and displacement correlates with purchasing power because it operates through money. The implication is

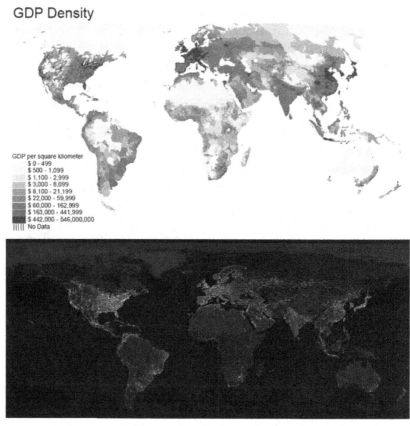

FIGURE 3 The geographies of money and technology

that modern technology is inextricably connected to what Martinez-Alier (2002) calls "ecological distribution conflicts." In other words, economic development and technological progress are intrinsically associated with issues of environmental justice. Conflicts over environmental justice are a central concern of the field of political ecology (Bryant and Bailey 1997; Peet, Robbins, and Watts 2011; Bryant 2015; Perreault, Bridge, and McCarthy 2015).

As Martinez-Alier (2002) shows, local victims of ecological distribution conflicts and environmental injustices all over the world have articulated myriad "valuation languages" and "idioms of resistance" that challenge the hegemonic assessments of market efficiency. The academic field of ecological economics, moreover, has developed several incontrovertible, quantitative indicators of the deleterious biophysical impacts of economic processes, but mainstream economists have generally remained unperturbed. Jointly, the environmental activists and critical academics today provide powerful signals that the economic models born in colonial Britain were not designed to promote a sustainable and equitable planet, yet no fundamental rethinking appears to be occurring within the mainstream discipline of economics. Although few would deny that current economic policies are inexorably generating rising inequalities, resource depletion, greenhouse gas emissions, and financial vulnerabilities, for a respectable economist to seriously question the design of the money artifact that is ultimately responsible for these processes would be as unthinkable as it would have been for a fourth-century Roman to question slavery or a fourteenth-century Easter Islander to question the construction of megalithic statues. In addition to submitting to the sheer inertia of conducting business as usual, of course, the tenacity of conventional economic thought reflects the implacable interests of powerful shareholders, executives, and politicians.

The innumerable examples of environmental injustice throughout the world provide a vast range of case studies in political ecology, illustrating how the tensions between corporate profit interests and the integrity of local ecosystems (including human communities) generate similar kinds of confrontations and political activism. Such tensions and conflicts can be classified and theorized in various ways, depending on whether we focus, for instance, on the discourse of the activists, the dynamics of the social movements, the type of corporations, the outcome of confrontations, or other parameters. For many years, however, my aim has rather been to grasp and articulate, in as succinct and widely applicable terms as possible, the common structural logic underlying such recurrent tensions between the imperatives of economics and the demands of justice and sustainability. My objective has been to distill the core concerns of political ecology and environmental justice, asking how the human reorganization of nature in various contexts has served to establish and reinforce social inequalities. These concerns have led me to develop an analytical framework for conceptualizing ecologically unequal exchange, which simultaneously provides a radically heterodox understanding of technology.

David Harvey (1996) also aims at revealing underlying mechanisms that generate the wide variety of environmental justice issues. To pursue this aim he is compelled to review (and frequently dismiss) a diverse range of contemporary discourses on the environment, suspecting that "the vast conceptual muddle and cacophony of discourses is far from innocent in

the reproduction of capitalism" (ibid. 173). He critically discusses a long
and rather arbitrary series of approaches to "how human beings have
valued their natural world," highlighting how unfeasible it is to reduce "a
wondrous multidimensional ecosystemic world of use values, of human
desires and needs, as well as of subjective meanings, to a common objective denominator," that is, *money* (ibid. 150–151). Harvey explores the
tension between the Marxist ambition to provide an abstract, universally
applicable theory of commodification, exploitation, and alienation, on
the one hand, and phenomenological approaches emphasizing specific
human experiences of place, on the other. He observes that "the increasing penetration of technological rationality, of commodification and
market values, and capital accumulation into social life (or into what
many writers, including Habermas, call 'the life world') together with
time-space compression, will provoke resistances that increasingly focus
on alternative constructions of place" (ibid. 302). Struggling to safeguard
his own Marxist approach at a time when universalizing grand narratives
were being discredited by the prevalent postmodern turn in social science,
he challenges "the postmodern death of justice," that is, the widespread
consensus that "there can be no universal conception of justice to which
we can appeal as a normative concept," but "only particular, competing,
fragmented, and heterogeneous conceptions of and discourses about justice which arise out of the particular situations of those involved" (ibid.
341–342).

Harvey deplores the "incredible vigor with which ruling interests have
sought to contain, shape, mystify, and muddy the contemporary debate
over nature and environment," for example "within discourses of 'ecological modernization,' ... 'sustainability,' 'green consumerism,' or the
commodification and sale of 'nature' as a cultural spectacle." In addressing the tendency of such discourses to obscure and confuse "the key
institutional and material practices that really matter for the perpetuation
of capitalist social and power relations" (ibid. 174–175), Harvey is more
intent on deconstructing discourses than on identifying those deleterious
and inequitable material practices. Although the 439 dense pages of his
book *Justice, Nature and the Geography of Difference* discuss almost
every other topic that might have some bearing on the relation between
nature and inequality, he makes no attempt to theorize unequal exchange
in terms of asymmetric transfers of physical components of nature, such
as energy, materials, or embodied land. This is tantalizing, as his aim is
precisely to establish a universalizing approach with which to challenge
the hegemony of exchange values:

The refusal to cast discussion in monetary terms ... reflects an intuitive or experiential understanding of how it is that seemingly fair market exchange always leads to the least privileged falling under the disciplinary sway of the more privileged and that *costs* are always visited on those who have to bow to money discipline while *benefits* always go to those who enjoy the personal authority conferred by wealth. There is an acute recognition within the environmental justice movement that the game is lost for the poor and marginalized as soon as any problem is cast in terms of the asymmetry of money exchange. (ibid. 388, emphases in original)

Climate Justice

In recent years, the pursuit of a structural account of global injustices pertaining to the biophysical environment has been particularly evident with regard to climate change. The greenhouse gas emissions that are largely responsible for the biogeochemical transformations of the Anthropocene are a quintessential illustration of the increasing entropy production generated by the global technological infrastructure fueled by fossil energy. They are thus an intrinsic aspect of what I have called "the thermodynamics of imperialism." The emissions have accelerated exponentially since the Industrial Revolution, and most conspicuously since World War II. Although the chemical composition of the atmosphere is a global phenomenon, the highly skewed distribution of emissions, their meteorological consequences, and the financial and technological capacity to cope with such consequences clearly establish that anthropogenic climate change is as inextricably connected with issues of global justice as the distribution of the technological infrastructure that is the source of those emissions.

In a study that demonstrates such connections in convincing detail, J. Timmons Roberts and Bradley C. Parks (2007) explain the problems plaguing international climate policy negotiations in terms of structural inequalities between the wealthier nations of the Global North and the less developed nations of the Global South.[5] They find that the available statistics on trade, GDP, and carbon dioxide emissions in different countries strongly support the theory of ecologically unequal exchange (ibid. 164–173), concluding that "ecologically unequal exchange is ... not just a perception by angry and irrational Third World politicians, leftist

[5] Roberts and Parks (2007: 164, 167) reiterate the suggestion of Bunker and others that the market prices of goods extracted from the South tend to be "underpriced" or "undervalued," which is a position that is both analytically flawed and unnecessary for the remainder of their argument.

scholars, and activists . . .; it is an observable physical reality" (ibid. 184). They also review evidence debunking the widespread notion that increasingly affluent economies finally tend to reduce their use of energy and materials, instead confirming that material-intensive production processes (and their emissions) tend to be shifted to poorer nations (ibid., 177, 183).[6] Although parts of their discussion of climate policy negotiations are marred by recurrent ambiguities regarding whether these structural inequalities are merely "beliefs," "worldviews," "ideas," and "perceptions" preventing "negotiators from reaching a shared social understanding of fairness from which to build the ambitious cooperative agreement needed to address climate change effectively" (ibid. 19, 23, 32–39), or whether they are indeed "a social reality" (ibid. 29), their argument as a whole is well substantiated, coherent, and commendably interdisciplinary.

The imperative of such an interdisciplinary outlook in discussing climate justice is illustrated by the constraints of a narrowly economic approach, as expounded by John Broome (2015). Assuming that the damages of climate change are "externalities," all translatable into monetary costs, and that continued economic growth and investment will benefit future generations, making them "better off than us," Broome at one point concludes that "climate change diminishes intergenerational inequality" (ibid. 16).[7] The logic of neoclassical economics is indeed impeccable, as in Lawrence Summers's infamous memorandum demonstrating the enhanced "efficiency" of shifting the most polluting industries to the poorest parts of the world, where the economic consequences of human illness would be less problematic (Summers 1991). Only a discipline as insulated from biophysical reality (and morality) as economics could reach as bizarre conclusions as these, yet they simply elaborate the alienated and fetishized logic of money. The conceptual myopia of mainstream economics is intrinsically inimical to considerations of

---

[6] The Assessment Report for the UNEP International Resource Panel emphasizes "the need to decouple economic growth and human well-being from ever-increasing consumption of natural resources," but establishes that the current tendency is precisely *opposite* to this imperative, as the "material intensity of the world economy has been *increasing* for the past decade," and globally, "more material per unit of GDP is now required" (Schandl et al. 2016: 14–16; emphasis added).

[7] Broome's (2015) solution to the problem of climate change is finally that governments should borrow more money for green investments. The currently soaring debts of most nations, and the ensuing financial crises, do not seem to concern him any more than the fact that the fantasies of financial institutions are incapable of alleviating the accelerating entropy production of our expanding use of energy.

environmental and climate justice. The logic may be impeccable, but if the premises and assumptions on which it is founded are fallacious – as are monetary reductionism and faith in perpetual growth – it can break down like a house of cards.

In this chapter I have argued that the global injustices of the Anthropocene can only be made visible by applying truly interdisciplinary perspectives that combine insights on both nature and society, while keeping the two aspects analytically distinct and refraining from the urge to reduce the natural to the social or *vice versa*. This approach makes it possible to rethink the ontology of modern technologies as societal strategies for the displacement of work as well as environmental loads, inherently contingent on asymmetric global transfers of biophysical resources and an accelerating production of entropy in the form, for instance, of carbon dioxide emissions. I have also indicated by means of which ontological assumptions these disturbing conditions of modern affluence are being kept out of view of mainstream economists, engineers, and people in general. However, the structural or "slow" violence (Nixon 2011) mystified by the lofty self-perceptions of liberal democracy is gradually revealing itself, in the form of obscene and growing global inequalities and the recognition that the affluent consumer society of the North cannot possibly be globalized (cf. Bauman 1998, 2011). Although the democracies of North America, Europe, and other wealthy parts of the world-system tend to take pride in their liberal virtues and respect for human rights, this self-congratulatory image is difficult to reconcile with their actual role in global society. We all need to uphold a virtuous moral identity, but it is becoming increasingly difficult for socialists and liberals alike to digest that their visions of a global welfare society are utopian, and that even those of us who are most intent to save the planet may count among its heaviest burdens.

4

The Money Game

In this chapter, I will draw in greater detail on Karl Polanyi's ([1944] 1957) analysis of the structural causes of human suffering that are immanent in the nineteenth-century idea of a free and self-regulating market, particularly for commoditized labor, and on David Graeber's (2011) inquiry into the 5,000-year history of such commoditization. I shall again trace the structural repression that they illuminate to the logic of general purpose money, viewed as a specific human artifact attributed with a particular social and ecological inertia. Such inertia of artifacts has been experienced by any player of a board game, and it is no coincidence that the discipline of economics has found extensive use for game theory.[1] The use of specific artifacts throughout human networks generates algorithmic regularities in social behavior that can be mathematically simulated. The current author makes no pretense to proficiency in such methods, but at a general level, I propose that the trajectories of human societies reflect the design of the artifacts that regulate their exchange relations, and that policies for transforming society thus cannot avoid considering how such artifacts are designed.

As I have previously conceded, to derive trajectories of social organization from regularities emanating from the features of given artifacts is an approach that suggests affinities with Actor-Network Theory (Latour 2005). The issue of how to approach the role of money artifacts in social

---

[1] By 2014, 11 game-theorists had won the Nobel Memorial Prize in Economic Sciences. An extensive popular literature on the activities on Wall Street and other financial markets also tends to represent investments and exchange as an obscure kind of game, in which success is more a matter of intuition than of following explicit rules ("Smith" 1967).

theory inevitably relates to the perception of money in socialist and Marxist understandings of modern economy. Some utopian socialists such as Robert Owen (1771–1858) realized that to build a society liberated from the injustices of early capitalism, it would need to abolish conventional money (North 2007: 43–50). Although profoundly aware of the repressive logic of money, Karl Marx dismissed the idea of abolishing money as idealistic, unless the capitalist relations of production were overthrown first. In Marx's worldview, money artifacts were reflections of capitalist property relations rather than their source. His insight that money is a fetishized representation of exchange relations did not prompt him to consider that the direction of influence between artifacts and relations could be reversed. A century and a half later, a leading proponent of Actor-Network Theory has explicitly challenged Marx's understanding of fetishism (Latour 2010). Although much of what is argued in Actor-Network Theory is problematic or unclear, to highlight the role of artifacts as generators of specific kinds of social relations is a significant contribution to social theory.

Viewing the economy as a game prompted and constrained by the features attributed to the pieces employed in that game raises the issue of which rules apply, whence they derive, and how they are communicated. For most people in modern nations, the only rules a player needs to know are that money is a remuneration for labor time or products sold, and that it can be used to purchase anything available through the market. The market is the economic aspect of society, the sum of all goods and services offered in exchange for money. If the money accessible to a given market actor is not sufficient to cover the market price of a commodity desired by that actor, additional money can be borrowed from (or rather created by) a bank. Specific rules regulate the rate at which the loan is to be repaid and the cost of borrowing it. These simple rules enable billions of humans to participate in an increasingly globalized economy. They are so general as to require no formal articulation or instruction. However, beyond these minimal rules there is an incredibly complex and constantly changing mass of regulations regarding the management of money flows between individuals, households, corporations, banks, authorities, and nations. These rules are the privileged knowledge of economists and are generally inaccessible or incomprehensible to people at large. Although the aspiration of economics is to articulate a set of regulations that would organize a perpetually viable economy, the history of recurrent financial crises clearly indicates that there are serious flaws in mainstream economic models (cf. Keen [2001] 2011; Mirowski 2013).

Karl Polanyi's Indictment of the Market

Toward the end of World War II, Karl Polanyi ([1944] 1957) summarized the turbulent history since the early nineteenth century in terms of a pervasive tension between the widespread liberal ambition to organize a self-regulating market, on the one hand, and the recurrent necessity to protect society against the various deleterious consequences of market principles, on the other.[2] In retrospect, Polanyi observed that the market system had an inherent tendency to degrade human laborers as well as land, neither of which were produced for sale but both of which were treated as commodities. At the outset of his book, he declares his thesis "that the idea of a self-adjusting market implied a stark utopia," putting faith in an institution that "could not exist for any length of time without annihilating the human and natural substance of society" (ibid., 3). The idea implied basing society, for the first time in history, primarily on *gain* as the explicit principle determining human relations. He shows that the wholesale subordination of society to market principles was as traumatic in early-nineteenth-century Europe as it has continued to be wherever the modern economy has engulfed previously nonmodern populations organized in other terms.[3]

Polanyi cites Aristotle's misgivings, more than two thousand years earlier, about production for gain as not "natural" to humans. Well versed in anthropology, Polanyi provides several examples from different societies of how humans have traditionally distinguished between local relations of reciprocity, on the one hand, and external markets, on the other. The Industrial Revolution in England, however, encouraged a fervent faith in the extension of market organization to include even what Polanyi calls the "fictitious" commodities of labor, land, and money. Much of the social history of nineteenth-century England can be understood as legislation and other initiatives for the protection of people against the pernicious consequences of transforming human lives into commodities. This countermovement aiming at the "self-protection of society" (ibid., 130) was prompted by the fact that the unregulated

---

[2] Writing during World War II, Polanyi believed that the globalized market economy born in the nineteenth century was rapidly declining. The resurgence of world trade after the war, and more emphatically with the neoliberal creed established in the 1980s, proved him wrong in terms of the tenacity of globalization, but not in terms of the fundamental tension between market and society that runs through his book.

[3] This part of his argument builds on the classical distinction between reciprocal gift-giving and the market exchange of commodities (Mauss [1925] 1990; cf. Gregory 1982).

market in early nineteenth-century England had the same implications for labor as it had for any other commodity: it depressed the price of labor to the absolute minimum required for survival. In contrast to the inclination of any premodern society, it was only expectable that this inexorable logic should tend to reduce early British proletarians to the verge of starvation.

But the notion of the self-regulating market was a threat also to land (i.e., nature) and to the supply of money. Polanyi observes that,

> if factory legislation and social laws were required to protect industrial man from the implications of the commodity fiction in regard to labor power, if land laws and agrarian tariffs were called into being by the necessity of protecting natural resources and the culture of the countryside against the implications of the commodity fiction in respect to them, it was equally true that central banking and the management of the monetary system were needed to keep manufactures and other productive enterprises safe from the harm involved in the commodity fiction as applied to money. (Ibid., 132)

To isolate land (i.e., an element of nature) and form a market out of it, Polanyi suggests, "was perhaps the weirdest of all undertakings of our ancestors" (ibid., 178). Beginning with the commercialization of the soil and the production of food and raw materials for industrial populations in Europe, the division of labor between rural and urban areas was extended to the planet. "With free trade," he writes, "the new and tremendous hazards of planetary interdependence sprang into being" (ibid., 181). "Once the great investments involved in the building of steamships and railroads came to fruition, whole continents were opened up and an avalanche of grain descended upon unhappy Europe" (ibid., 182). As Polanyi notes, the perils of relying on a global market for basic subsistence became obvious with World War I, highlighting the necessity of protectionist measures to provide nations with some measure of agricultural self-sufficiency. European nations could protect themselves against the repercussions of free international trade (e.g., through the introduction of corn laws in Central Europe), but the revolts against imperialism in the colonies should be understood as attempts to "shelter themselves from the social dislocations caused by European trade policies" (ibid., 183). Even if the notion of free trade and a global self-regulating market might appear beneficial from the perspective of a European businessman, its repercussions on other continents were very far from benevolent. The market freedom celebrated by liberals to this day remains a zero-sum game.

World trade, finally, could only be relied upon with a system of international commodity money, as national token (or "artificial") money

cannot circulate outside the nation.[4] The gold standard was thus introduced in the beginning of the nineteenth century to stabilize international
exchange, but the recurrent problem since then has been the utopian
project of maintaining a viable calibration between national currencies
and an international gold standard. Given that volumes of production
and exchange in different nations tend to change independently of each
other, and that volumes of national currencies are internally calibrated
with such volumes of economic activity to avoid deflation and domestic
unemployment, the relation between the two kinds of money is inevitably
strained. To overproduce token money results in inflation. The incentive
for nations to issue more token money than can be backed by their
deposits of gold has proven pervasive, and world economic history over
the past two centuries has been punctuated by national relinquishments
and resumptions of the connection to gold.

These complex challenges of national monetary policy required central
banking, and Polanyi observes that a nation's economic identity is "vested
in the central bank" (ibid., 205). World trade through much of the
nineteenth century was dominated by the gold standard maintained by
the Bank of England. In issuing large quantities of token money to finance
its military efforts during World War I, however, England disregarded the
gold standard to such an extent that in 1931, after a brief attempt to
resume it, the country permanently dropped the connection between its
currency and its deposits of gold. The United States followed its example
two years later, as did France and some other countries by the mid-1930s.
The global economic depression that began in 1929 had already dramatically reduced world trade, but Polanyi concludes that the "final failure of
the gold standard was the final failure of market economy" (ibid., 200).[5]
International debts were repudiated and the principles of economic liberalism, which had been celebrated in the 1920s, were largely rejected.

Paradoxically, to make the idea of a self-regulating market feasible,
economic liberalism has required "an enormous increase in the adminis-
trative functions of the state" (ibid., 139). As Polanyi observes, "[T]here
was nothing natural about *laissez-faire*; free markets could never have
come into being merely by allowing things to take their course."

---

[4] Polanyi explains that national token money was developed to alleviate the chronic scarcity
of money in relation to the domestic expansion of production and trade, which would
otherwise lead to deflation.
[5] The mobility of money, however, in the years of the depression compensated somewhat for
the protectionist barriers that reduced the flow of people and goods.

Internationally, the vision of a free world market required rigorous political and often military enforcement. Whereas liberal economists tend to blame the breakdown of nineteenth-century globalization on imperialist rivalry gradually gathering strength and leading to World War I, Polanyi dismisses such an account as myth, arguing instead that it was the "economic earthquake" of increasing international trade that prompted protectionist measures and nationalism. This was evident already during the depression of 1873 – 1886, during which Western nations protected themselves from unemployment and instability "with the help of central banks and customs tariffs, supplemented by migration laws" (ibid., 217). Economic imperialism was simply the concomitant struggle of these nations to extend their trade into politically unprotected markets.

Political and economic crises, Polanyi argues, are generated by the continuous tension between the utopian vision of a free market and the several policies for self-protection to which nations are compelled to resort. It is regarding the nature of this tension that Polanyi disagrees with liberal economists. While Polanyi is convinced "that the inherent absurdity of the idea of a self-regulating market system would have eventually destroyed society," liberals accuse the various counterreactions to that idea of "having wrecked a great initiative" (ibid., 145). The depression of the 1930s signified the disintegration of the world economy and "the almost unbelievable fact that a civilization was being disrupted by the blind action of soulless institutions the only purpose of which was the automatic increase of material welfare" (ibid., 219). "The stupendous industrial achievements of market economy had been bought at the price of great harm to the substance of society" (ibid., 186). The bottom line of Polanyi's argument, however, is that the contradictions generated by liberal capitalism are the common denominator of various fascist movements through history, including those that ignited World War II. "Fascism, like socialism," Polanyi asserts, "was rooted in a market society that refused to function" (ibid., 239).

Taming and Unleashing the Market

In hindsight, we know that the international gold standard and conditions for free trade were quickly restored at Bretton Woods after World War II,[6] and that the American dollar remained pegged to gold for a quarter of

---

[6] The agreement at Bretton Woods was reached the same year (1944) that Polanyi published *The Great Transformation.*

a century until 1971, when President Nixon had spent so much money on the war in Vietnam that he was compelled, again, to abandon the connection. During those two-and-a-half postwar decades, the industrial economies of the world enjoyed unprecedented expansion, and affluent European nations such as Sweden were able to offer models of market society tempered by a mild form of socialism. In the wake of the oil crises and economic instability of the 1970s, however, leading politicians in England and the United States launched a neoliberal program that dismantled much social security and other forms of political intervention in the operation of the market. The decades following this turn have brought deepening economic inequalities between and within nations as well as alarming environmental degradation worldwide. Moreover, the major financial crisis beginning in 2008 has frequently been explained in terms of the expansion of "financialization" following the abandonment of the gold standard 37 years previously. Although the gold standard for almost 170 years had remained the foundation of a utopian vision of a self-regulated world market, neither its maintenance nor its absence has enhanced global financial equity or sustainability. In the nineteenth century as well as during the postwar decades, free international trade, enhanced by the gold standard, has enriched some nations at the expense of others. Financial stability is not to be confused with equality.

The tension between proponents of market principles and societal self-protection that Polanyi in the 1940s identifies through history recurs in modern times and at successive geographical scales. The formation of national markets by mercantilist European states in the fifteenth and sixteenth centuries required the subjugation of local particularism and protectionist towns and municipalities; five centuries later the European Union extended similar visions of a common market to the continental scale, reducing the autonomy of the constituent nations. In 2016, however, a majority of British voters decided that their country should leave the European common market, and it seemed possible that other nations would follow. Such movements for national protectionism continue to be viewed by liberals as regressive and a threat to world peace. Even the social democrats who for most of the twentieth century protected the population of Sweden from the worst ravages of the market economy now tend to adhere to the neoliberal approach to globalization. It should be observed that the political project of social democracy, rather than a realistic model for all nations, has been feasible only in affluent nations capable of displacing market-generated poverty and inequalities to others. The social condition of modernity cannot be globalized.[7]

Polanyi asserts that "socialism is, essentially, the tendency inherent in an industrial civilization to transcend the self-regulating market by consciously subordinating it to a democratic society" (ibid., 234). At the time of his writing, and up to the 1970s, it was still reasonable to believe in the prospects of an industrialized but socially domesticated market economy like that of Sweden under social democracy. He was thus able to predict, for instance, that "industrial civilization will continue to exist when the utopian experiment of a self-regulating market will be no more than a memory" (ibid., 250).[8] But in hindsight, Polanyi appears to have overlooked the extent to which national self-protection is feasible only if damages to people and the environment can be displaced to other nations. The issue boils down to the delineation of society. If the damages generated by market principles cannot be resolved but merely displaced, the question is ultimately how a *world* society shall be able to protect itself from the market.

Neoliberalism is a renewed and more relentless form of market fundamentalism provoked into existence by the failures of capital accumulation in the 1970s. The neoliberal transformations of Sweden and other European welfare states since the 1980s suggest a parallel, at the national scale, to the failure of local experiments with "utopian socialism" such as that of Robert Owen a century and a half earlier.[9] Again, the contradiction between economic liberalization and societal self-protection can be identified at separate geographical scales and separate historical times. Considering the currently widening global inequalities, the escalating environmental problems generated by world capitalism, and the increasingly vulnerable global financial system, we have every reason to revisit Karl Polanyi's indictment of the liberal idea of the self-regulating market. These three incontrovertibly unsustainable trajectories of globalization correspond to the deterioration of precisely those three factors of

---

[7] As David Graeber (2011: 355) observes, "[W]e could no more have a universal world market than we could have a system in which everyone who wasn't a capitalist was somehow able to become a respectable, regularly paid wage laborer with access to adequate dental care. A world like that has never existed and never could exist."

[8] The modified market system of the future that he advocates will not be self-regulating, he suggests, as it will not comprise labor, land, and money.

[9] Robert Owen's vision of socialism, generally dismissed by posterity as utopian, is treated by Polanyi with the utmost respect (ibid., 258A).

production that Polanyi classified as "fictitious commodities": labor, land, and money.

Polanyi identifies many of the pernicious social consequences of market logic. He shows how untamed industrial capitalism in nineteenth-century England tended to make workers "physically dehumanized" and "the owning classes ... morally degraded" (ibid., 102). Until the workers had been made dependent on "the mill of the market which ground the lives of the people," popular democracy was unanimously understood as "a danger to capitalism" (ibid., 226). If labor is indeed supposed to find its price on the market, Polanyi observes, it should be expected "to be almost continually on strike" (ibid., 230–231). Such absurdities are only "logical inference from the commodity theory of labor."

The Logic of Global Inequalities as Emergent from Money

We have every reason to agree with Polanyi that the history of the past two centuries has been dominated by a tension between a blind logic of some sort and the various forces and movements that have attempted to curb its most deleterious repercussions. The crucial issue, however, is how this blind logic is conceptualized. Karl Marx framed it in terms of a mode of production referred to as *capitalism*. Polanyi rarely refers to capitalism but prefers to identify the problem as "the idea of the self-regulating market."[10] Both Marx and Polanyi are acutely aware of the inequities and human degradation that are the flip side of economic growth and technological progress since the Industrial Revolution, but both may have underplayed the root of the societal inertia that disturbs them. To be sure, it is easy to identify human advocates of this inertia, whether defined as a class of capitalists or a school of liberal economic theorists, but, given the pervasive failures to curb its destructive consequences through politics or economic discourse, we have reason to look elsewhere for its final source.

If the Actor-Network theorists are correct in deriving significant aspects of human social relations from features of the artifacts that mediate those relations, we should find ourselves seriously revisiting St. Paul's observation that "the love of money is the root of all evil." What

---

[10] Fernand Braudel has distinguished between markets as the use of money to convert one commodity into another (C-M-C'), on the one hand, and capitalism (M-C-M') as the use of commodities to augment one's supply of money, on the other (cf. Graeber 2011: 260).

Marx expounded as the logic of capital, and Polanyi as the idea of the market, can be viewed as the societal implications of the inertia of general purpose money. The incentives driving the behavior of capitalists and other market actors can be derived from the possibilities and imperatives that are immanent in the very artifact of conventional money. The concept of artifact as used here encompasses items manufactured by humans as well as the features attributed to them by their users. These features are ultimately ideas, but when the items with which they are invested circulate in society according to the logic determined by those ideas, they form regularities and trajectories of their own that are external to the individual human mind and may remain largely opaque to the participants. A relevant illustration of such trajectories determined by the features of artifacts is the game of chess. The potential mobility of a queen is distinctly different from that of a knight or a pawn. The possible trajectories of the game are determined by the attributes of the different pieces as defined by its rules. Transferred to the game of market capitalism, a corollary of this perspective is that its societal logic is determined by the attributes of money.[11]

It no doubt requires a shift in habits of thought to attribute responsibility for human social relations to features of the artifacts that are employed in mediating those relations, but this may indeed be what is required of us in the current global impasse. It means finally fathoming the implications of Marx's insights on *fetishism*. We need to understand that the market economy, capitalism, and even the technological progress of the Industrial Revolution are all ultimately products of our preoccupation with general purpose money. Even in Polanyi's narrative, such insights are never far away. He understands that a market economy "assumes the presence of money, which functions as purchasing power in the hands of its owners" (ibid., 68), that eighteenth-century national markets "presupposed the almost general use of money" (ibid., 115), and that the Owenites' attempts to transform the social logic of capitalism required "introducing a currency of its own" (ibid., 169).

At some level, Polanyi is also aware that mechanization is geared to the commoditization of labor, land, and money. He writes that we cannot

---

[11] Every now and then, Polanyi casually refers to the market economy as a game (ibid., 160, 188, 197). From several different perspectives, it can indeed be shown to have features of a zero-sum game. The market economy thus not only displaces work and environmental loads (Hornborg 2013a), it redistributes risk, freedom, mobility (Bauman 1998), and even honor (Graeber 2011: 175).

fully grasp the nature of the market economy unless we realize "the impact of the machine on a commercial society" (ibid., 40) and that "machine production in a commercial society involves ... no less a transformation than that of the natural and human substance of society into commodities" (ibid., 42). He observes that the nineteenth century, "whether hailing the fact as the apex of civilization or deploring it as a cancerous growth," naïvely imagined that the market economy was a "natural outcome" rather than the "artificial" result of policies designed "to meet a situation which was created by the no less artificial phenomenon of the machine" (ibid., 57). His recognition that the Industrial Revolution was inextricably interconnected with the market economy is evident in his conclusion that "either machines had to be demolished, as the Luddites had tried to do, or a regular labor market had to be created" (ibid., 81). According to Polanyi, the immediate self-interest of workers "destined them to become the protectors of society against the intrinsic dangers of a machine civilization" (ibid., 101). The Owenite movement, he asserts, represented the desire "to discover a form of existence which would make man master of the machine" (ibid., 167). The relation between machine technology, markets, and money to which he refers is significant but, as I shall show, widely misunderstood.

I have for many years tried to communicate an understanding of mechanization and technology that would turn Polanyi's view of the relation between machines and markets on its head. Not only does industrialized production prompt factory owners to commoditize labor, land, and money, but industrial technology is in itself a *product* of such commoditization. The Industrial Revolution in England was made possible by the global flows of embodied labor and land. Industrial technology is an enhancement of the productivity of local labor through the appropriation, by means of asymmetric exchange, of commoditized human time and natural space. Ultimately, all commoditization, and the new forms of asymmetric exchange associated with industrialization, are generated by the very idea of general purpose money. In the same sense that a biological organism is the manifestation of its metabolic exchanges with its environment, the technological infrastructure of industrial society is the tangible manifestation of global relations of resource exchange, orchestrated by money. Money and technology are thus recursively and inextricably intertwined. To approach economic growth and technological progress as separable indices of development is a pervasive illusion that continues to obstruct our grasp of contemporary concerns regarding global justice, sustainability, and financial resilience.

Polanyi recognized the *immoral* implications of market logic as a pervasive trajectory of liberal ideals of self-regulation. The relation between economic exchange, money, and morality is complex and has been extensively discussed by theologians, philosophers, sociologists, and anthropologists (Bloch and Parry 1989). My point here is that the establishment of neoclassical economics in the final third of the nineteenth century entailed an abandonment of moral concerns. Prior schools of economic thought had applied various criteria of morality to exchange, such as thrift, land management, or justice. Their moral arguments had in various ways been connected to concerns with the material substance of exchange, whether advocating restraint in exporting precious metals, care for agricultural soils, or a labor theory of value. The turn to an exclusive preoccupation with market equilibrium in Victorian Britain implied the simultaneous and interrelated abandonment of morality and materiality as concerns of economics. The only kinds of moral assessment relevant from now on were concerns with the extent to which market mechanisms had been permitted to operate freely. To neoclassical economists the moral notion of "unequal exchange" thus denotes market power (such as monopoly) rather than an asymmetric exchange of resources, as identified by heterodox approaches (cf. Emmanuel 1972; Dorninger and Hornborg 2015; Hornborg and Martinez-Alier 2016). As all other possible measures of commodity flows (such as embodied labor, land, energy, or materials) are made invisible by the exclusive concern with monetary exchange-values (utility), any voluntary market transaction is fair *by definition*. Free market prices axiomatically guarantee reciprocal exchange because concerns with reciprocity are delegated to monetary assessments. In this view, by implication, a net transfer of material resources to core sectors of world society should somehow – and incongruously – be irrelevant to their prospects of accumulating technological infrastructure. In a nutshell, this indicates how money and technology are intertwined. The establishment of neoclassical economics protected capital accumulation from moral objections based on the asymmetry of biophysical resource flows.

Polanyi's point that the commoditization of labor and land, if left to a self-regulating market, would severely degrade both people and nature is both an analytical deduction from economic theory and an empirical inference from the history of early industrialism. Although a major theorist of money, he did not adopt the Marxian view that monetary exchange-values tend to obscure the transfer of more substantial values. However, if he had wanted to demonstrate that unrestrained market exchange will tend to deprive both labor and land of their *physical* capacity for regeneration, it would have been completely in line with his conclusions. It is remarkable overall that the voluminous and profound reflections of economic anthropologists on money and exchange so rarely go beyond the cultural semiotics of exchange to consider how that semiotic veil orchestrates material transfers. My guess is that anthropological training is so exclusively devoted to the mental (cognitive, cultural, symbolic) aspects of human existence that material aspects remain inaccessible both theoretically and methodologically. The only example of such interdisciplinarity in economic anthropology that I can think of is Maurice Godelier's (1969) identification, 50 years ago, of the unequal exchange of embodied labor time between two tribes of New Guinea. Regardless of Godelier's conclusions, such attempts to unearth biophysical flows veiled by cultural institutions of reciprocal exchange are unheard of among anthropologists today.

Money and Morality

David Graeber's (2011) reflections on the history of money and debt provide a very significant complement to the foundational contributions to economic anthropology by Mauss ([1925] 1990) and Polanyi. He perspicaciously undermines our familiarity with monetary debt by illuminating its historical origins and contrasting it with patterns of exchange and obligation in a wide variety of cultural contexts. The argument is highly pertinent to the present discussion of destructive tendencies inherent in market exchange, its ambiguous relations to morality, and the identification of financial crisis as a manifestation of its potential for structural repression. Modern money, Graeber notes, is debt. Long before debt could be calculated in the unnegotiable, mathematical terms of money, interpersonal obligations were embedded in compelling moral imperatives, but Graeber's ambition is to explore how the precise, quantitative language of the market transforms our sense of reciprocity, morality, and justice – that is, how the autonomous, mechanical rationality of cost–benefit calculation (and its frequently deadly repercussions) has emerged out of a moral sociality grounded in interpersonal relations.[12]

---

[12] This is the perennial concern of economic anthropology, epitomized in Polanyi's observations on the disembeddedness of the modern market. In delegating the logic of social relations to the fetishism of money and market mechanisms, it is as if humans can dissociate themselves from, and relinquish responsibility for, the suffering they impose

Graeber proposes that there have been historical oscillations between periods dominated by virtual or credit money, on the one hand, and commodity money made from bullion, on the other. He associates periods dominated by commodity money (800 BC–AD 600, 1450–1971) with relatively more warfare, plunder, and slavery, while periods of credit money (3500–800 BC, AD 600–1450, 1971–?) are associated with relatively more social peace and trust. The reason why a predominance of precious metals correlates with periods of widespread war and plunder, says Graeber, is that, unlike credit arrangements, metals can be physically stolen. In ancient Greece, coinage made from bullion inspired the notion of a substance that could be converted into everything, which is precisely the modern concept of money.[13] Although the empirical historical correlation between commodity money and warfare (and between credit money and peace) is not quite as neat as Graeber suggests (cf. Weatherford 1997; Davies 2002; Ferguson 2008), his argument that coins made of bullion were widely used to pay for armies and the capture of slaves (to mine more bullion) is convincing. As he demonstrates, however, slavery can be a consequence of debt as much as violent conquest, and the distinction between Roman slavery and feudal debt peonage is thus ultimately as ambiguous as that between slavery and wage labor. Graeber notes that most slaves exported from the Bight of Biafra in the 1760s had become slaves through debt servitude, and that "commercial economies had already been extracting slaves from human economies for thousands of years," finally asking if the practice is "actually constitutive of civilization itself?" (ibid., 163). Money and markets indeed appear to be institutions for transforming persons into instruments of production. Wage labor, Graeber suggests, is "the renting of our freedom in the same way that slavery can be conceived as its sale" (ibid., 206).

Graeber notes that most people simultaneously hold the two contradictory convictions that to pay back one's loan is a matter of morality, while moneylenders are evil. The financial crisis beginning in 2008 exposed a "host of new, ultra-sophisticated financial innovations" that had been devised by bankers but that were intentionally made

---

on other people and the planet. "Any system that reduces the world to numbers," Graeber concludes, "can only be held in place by weapons" (ibid., 386). A debt, he finally reminds us, is "a promise corrupted by both math and violence" (ibid., 391).

[13] Graeber insightfully reviews the theological dimensions of money, including the denunciation of usury in Islamic and early Christian thought (cf. Le Goff 2012). The long-standing concerns over the conundrum of money are graphically reflected in Shakespeare's *The Merchant of Venice* (cf. Macfarlane 1985).

impossible to fathom for the layman, as "nothing more than very elaborate scams" (ibid., 15). Such scams are also being perpetrated at the
international level. Some nations (most conspicuously the United States)
can count on an indefinite extension of international credit that assumes
the appearance of tribute, rather than foreign loans, while other nations
are punished with austerity measures and coerced into debt repayment
and poverty. "American imperial power," Graeber says, "is based on a
debt that will never – can never – be repaid" (ibid., 367). It is a war debt
and amounts to a tax paid to the United States by the whole world. He
refers to it as "debt imperialism." Graeber's understanding of world
capitalism is well summed up in the assertion that,

> it is a gigantic financial apparatus of credit and debt that operates ... to pump
> more and more labor out of just about everyone with whom it comes into contact,
> and as a result produces an endlessly expanding volume of material goods. It does
> so not just by moral compulsion, but above all by using moral compulsion to
> mobilize sheer physical force. (ibid., 346)

A significant aspect of Graeber's world history of money and debt is the
recognition that an underlying tendency toward a fetishized, money-
mediated repression of other people is evident over such a long period
and in so diverse cultural contexts. He notes that "almost all elements of
financial apparatus that we've come to associate with capitalism ... came
into being not only before the science of economics ... but also before the
rise of factories, and wage labor itself" (ibid., 345). This confirms the point
that I made previously about money being a requisite of the Industrial
Revolution and the very phenomenon of modern technology. It is striking,
however, how the ontology of technology in mainstream thought – even
among the most perspicacious critics of capitalism – remains sequestered
from and uncontaminated by our most profound insights on social relations of exploitation. Although highly relevant to asymmetric flows of
embodied labor, and although he occasionally mentions the degradation
of the planet, Graeber's analysis never extends into serious consideration
of material phenomena such as technology, energy, or ecology.

A central merit of Graeber's study is its recurrent acknowledgment of
the magical ontology of money. Money, he says, "has no essence" (ibid.,
372). Even Aristotle saw that money, even gold, is just a convention.
Faust (i.e., Goethe) and Keynes were equally aware that it does not matter
how much money you print, as long as people trust it. To consider it
fraudulent that bankers create money out of nothing is a sign of materialism and a naïve belief in money as essence – an indication, indeed, that we

have been persuaded by their magic. This is the nature of the bubble currently enveloping the world. The players of the money game know all too well that it "can all come tumbling down" ("Smith" 1967: 238). Sooner or later it must. But the ontological dilemma we seem unable to digest is that what Keynes called "organic propositions" (the truth of which depends on *beliefs*) can organize physical reality. The contemporary extraction of unconventional fossil fuels in North America would not be feasible without the financial fantasies of Wall Street. This means that we are allowing the human fantasy of money to magically destroy the planet. To thus fully acknowledge the interpenetration of human fantasy and physical reality – a modern form of magic that remains difficult for post-Enlightenment thought to digest – is our only chance of surviving the Anthropocene. In this specific sense, the Anthropocene does pose a challenge to Enlightenment dichotomies such as subject-object and society-nature. Enlightenment rationality dismissed premodern, magical notions about the capacity of subjectivity to intervene in objective conditions, but is now compelled to embrace that such intervention can be accounted for in perfectly rational terms.

There is something profoundly emancipatory in the insight that money is ultimately an idea and its current form merely a convention that could, in principle, be redesigned. Many attempts have been made over the years (cf. North 2007), but all have ultimately failed, and for similar reasons. Rather than revolutionary movements challenging hegemonic power centers such as the Bank of England or Wall Street, the introduction of new forms of currency needs to be anchored in the political intentions of democratically elected authorities. It is through democracy that the rules of the game might be rewritten, the pieces redesigned, and the players liberated from their delusions. But for the incentives to be strong enough to so thoroughly rewrite the rules, the current money game will have to more decisively expose its inadequacies.

5

Anticipating Degrowth

Recognizing the many deleterious repercussions of economic growth, many critics of these tendencies have advocated so-called *degrowth* (Latouche 2009; D'Alisa et al. 2015; Kallis 2018). Giorgos Kallis (2018: 4–8) shows how Serge Latouche's concept of degrowth in 1972 represented the convergence of perspectives from several sources including the economist Nicholas Georgescu-Roegen, the philosophers André Gorz and Ivan Illich, and the economic anthropologists Karl Polanyi, Marcel Mauss, and Marshall Sahlins. The relation between visions of degrowth and Marxism, however, remains ambiguous. Although "capitalism" is frequently identified as the main target of critique, proponents of degrowth rarely clarify how they conceptualize technological progress, unequal exchange, surplus production, or the exploitation of labor and nature.

I am not aware of any politician anywhere who is seriously advocating degrowth. This is not difficult to understand, given the conventional trust in economic growth as the source of solutions to most of the problems that trouble us. The aims and demands of most social groups seem to be possible to meet by allocating more money to them. The extent to which this belief in monetary solutions is illusory is highlighted by contemporary efforts to persuade us that even the adverse consequences of economic growth – such as the many signs of globally deteriorating ecological conditions – can be remedied with more growth (Asafu-Adjaye et al. 2015). It would be superfluous here to demonstrate how contradictory this "ecomodernist" ideology really is; it will suffice to repeat Kenneth Boulding's famous observation that "anyone who believes exponential growth can go on forever in a finite world is either a madman or an economist." Nevertheless, the point here is that any presumptive politician naïve or honest enough to advocate degrowth is unlikely to have a future in politics. Economic growth continues to be the foundation of most people's hopes and aspirations. It thus seems unlikely that a policy of *intentional* degrowth will be compatible with democracy in the foreseeable future. Unfortunately, unless pressed by severe and acute crisis, a democracy cannot be expected to *decide* to degrow.

This should not be understood as a dismissal of the discourse on degrowth, however. An *involuntary* contraction of economic activity is a highly realistic scenario for many nations in the coming decades, regardless of popular opinion. As illustrated by historical as well as recent experiences of financial crises, periods of economic decline are difficult to predict and even more difficult to evade. There is every reason to share Queen Elizabeth's concerns about the failure of the economists to predict the crisis of 2008 (Besley and Hennessy 2009). The crisis highlighted the inability of conventional economic models to grasp the total logic of human economies, here understood as immensely complex systems of human artifacts interacting with global social, political, cultural, and ecological aspects of our existence. Although the image of politicians in control of national destinies is widely projected in public discourse, historians know that it is largely illusory. Financial crises remind us of our collective vulnerability, and recent crises remind us that the extent of our vulnerability has increased with the continued integration of the global economy. Economic globalization increases our global interdependence, which means that a major disruption of the global economic system will have more severe consequences even for our individual subsistence needs.

The generally unanticipated financial crisis of 2008 allows us to conclude that the task of assessing the probability of economic decline ought not to be left to conventional economists. It poses an interdisciplinary challenge extending far beyond the discourse of mainstream economics. As Tim Jackson (2009: 185) concludes, "[T]he clearest message from the financial crisis of 2008 is that our current model of economic success is fundamentally flawed." Although entire libraries can be filled with publications debating economic theory and policy, virtually all this discourse is conducted within the constraints established by the nineteenth-century assumption of general purpose money. This means that the entire range of modern economic thought is restricted to conditions set by a specific (and historically recent) human artifact. Over the past century, we have seen the logic of money inflate and explode various kinds of financial bubbles, but the most fundamental bubble of them all is the belief in money.

As many commentators have predicted regarding the exponentially expanding debt of the United States and several other affluent nations, the end of the current trajectory will be calamitous, to say the least. Even if financial crises can be temporarily averted by gaining access to expanding credit, this will only inflate the bubble even more and postpone an inevitable and ever more disastrous collapse.

Traditional Marxists would understand economic crisis as inherent in the logic of capitalism: the chronic incapacity of the economy to create enough purchasing power to realize the surplus value embodied in commodities and, finally, the long-term tendency of the rate of profit to fall, leading in the end to a revolution and the creation of a postcapitalist society. But not even the heterodox economic models of Marxism are interdisciplinary in the sense that they acknowledge the role of physical factors such as EROEI, or net energy (Hall and Klitgaard 2011). This is a topic to which I will return in Chapters 8 and 9.

Considering how detached both orthodox and heterodox economic models are from the physical reality in which the economy is embedded, it seems unwarranted to expect economists to provide adequate accounts of the causes of economic crises, let alone feasible remedies. The complex social logic generated by the artifact of modern money is obviously intricate enough to preoccupy millions of hard-working, meticulously trained, and finely wired minds around the world, but precisely because they are thus wired these minds are not likely to adopt a detached perspective on how this peculiar human artifact is responsible for the disastrous course of global society. Few economists have given a thought to how maladapted the idea of general purpose money is to the physical conditions of the biosphere, which it increasingly dominates (Georgescu-Roegen 1971; Daly 1996). Unfortunately, those few dissidents have had very little influence on mainstream, neoclassical economic concerns, which continue to focus on the infinitely complex logic of market equilibrium and the management of money.

Money Fetishism

To understand this societal preoccupation with the operation of the specific artifact of money, rather than with our chances of substantially redesigning it to better harmonize with biophysical conditions, we need to return to the concept of *fetishism*. It was first applied to political economy in the mid-nineteenth century by Karl Marx, who observed that the Europeans of his time were as enthralled by their own artifacts (commodities and money) as were the worshippers of idols encountered by Portuguese merchants along the West Coast of Africa (cf. Morris and Leonard 2017). This detached view of everyday economy as a peculiar cultural game, largely determined by the properties attributed to a particular human artifact, represents a paradigm shift in our understanding not only of the organization of human societies but also of its implications for the ecological systems with which they are intertwined. The phenomenon of economic growth is a socio-ecological process the full explanation of which requires a detached and defamiliarizing perspective on taken-for-granted elements of everyday life, most essentially money, as well as a truly interdisciplinary approach to the conditions for capital accumulation. Mainstream, neoclassical economic theory has shown no interest in either cross-cultural defamiliarization or interdisciplinarity, but continues to present itself as the exclusive authority on growth and everything else pertaining to the operation of human economies.[1] Its models are complex and inaccessible to the layman, but are so entrenched in the fetishized logic of the money artifact that, predictably, they provide scant guidance on how to deal with the global problems of sustainability generated by that very artifact. Not only are competent concerns with ecology and global justice dispelled beyond the horizons of economics, but as recent events have demonstrated, even the conditions of financial resilience remain a mystery for the field. Interdisciplinary attempts to understand how and why the world economy is generating planetary crises, escalating inequalities, and ominous financial instability are systematically barred from official knowledge production and decision making by the guardians of the discipline of economics, which regards the operation of money as its own privileged territory and any external critique of its fetishized models as indicative of ignorance. It is no exaggeration to assert that the global future of humankind hinges on its capacity to transcend the ideological constraints of such money fetishism.

Modern money boils down to the idea that all things and services are interchangeable. This means that previously unthinkable conversions and transformations are made possible, bringing together materials and designs from around the globe. Money and global trade are thus sources

---

[1] An example of this conceit is the ambition to deal with ecological degradation through the lens of so-called environmental economics, a subfield of neoclassical economics focusing on people's willingness to *pay* for ecological considerations but requiring no competence in ecology or other natural science. As Tim Jackson (2009: 123) has observed, economics is "ecologically illiterate."

of unprecedented creativity, as most prominently evident in the development of technology over the past three centuries. A paradigmatic example is the Industrial Revolution. The early industrial districts of England in the late eighteenth century were able to combine cotton fiber from America, iron from Sweden, domestic coal, and several other inputs – extracted, transported, and transformed by vast armies of commodified labor – into textiles and other merchandise destined for the same globalized market, thus recursively providing access to increasing quantities of such inputs and other commodities. The idea and artifact of money has catalyzed global socio-ecological processes of economic growth that, by and large, have been celebrated as progress. Although justified, the various kinds of criticism that have been directed at the cataclysmic ecological, political, social, and cultural repercussions of these processes have never seriously threatened their continuation. The inertia of money and economic globalization has proven impossible to restrain, whichever national or local policies have been launched to curb it. Even international attempts to control its impacts on planetary carbon cycles have been futile. This futility of all reservations against economic growth and its global consequences is not fortuitous, for they very rarely address the peculiar character of the money artifact that is the core driver of these processes.

Of the many far-reaching implications of money for social life, ecology, politics, and culture, its most insidious feature is its capacity to equate, in terms of exchange-value, commodities that have quite different qualities, embody very different quantities of resources, and possess highly divergent productive potential. The seemingly objective representation of very different goods and services as equivalent makes it possible to systematically appropriate increasing quantities of resources through the market *without violating the appearance of reciprocity*. For Marxists, this is a requisite of capitalism, which obscures the discrepancy between the exchange-value and "use-value" of labor-power. Although the productive resource that most concerned Marx was labor-power, perspectives from ecological economics have illuminated how other biophysical resources, such as embodied energy, matter, and land, may also be asymmetrically transferred through seemingly reciprocal market transactions (Dorninger and Hornborg 2015). Such asymmetric market transfers make it possible to accumulate productive infrastructure like the textile factories that inaugurated the Industrial Revolution. Capital accumulation is an emergent property of money, as money makes it possible for the market to reward resource dissipation with more resources to dissipate. Again, this is not to say that the asymmetrically transferred resources have a higher "value" than their market price, but that the infrastructure in which they are incorporated can be used to produce increasing volumes of exchange-values (i.e., monetary profits). Accumulation and profits are thus contingent on asymmetric transfers of biophysical resources.

The artifact of money generates profit incentives[2] and an increasingly globalized trade of embodied labor, energy, materials, and land. Among the ecological consequences of economic globalization are an accelerating resource use, in part because of the expansion of long-distance transports (and the additional production and consumption that they permit), in part because globalization discourages recycling. A primary incentive driving globalization is the urge to exploit international differences in wages and the market prices of other resources. The lowering of costs made possible by shifting production to low-wage areas of the world economy increases demand and consumption in wealthier areas. Rather than employ high-wage labor in recycling, moreover, it is more rational for wealthy countries to consume increasing amounts of fresh resources from low-wage countries.[3] Nor is it feasible for resources to be returned to distant zones of extraction, illustrating what John Bellamy Foster (2000) has identified as the "metabolic rift" that is part and parcel of capitalism. Even climate change must be viewed as a consequence of accelerating resource use in transports, production, and consumption, and thus geared to globalization and ultimately the logic of money. However, these ecological dilemmas have not discouraged the proponents of free trade from asserting that the world market is the best road to sustainability.

The Nation versus the Globe?

Moreover, globalization continues to generate increases in inequality (Alvaredo et al. 2017) and various problematic sociopolitical and cultural consequences of economic polarization. Because mainstream neoliberalism advocates economic policies that dismantle social welfare, outsource employment opportunities to countries with lower wages, and exacerbate inequalities within and between nations, it tends to produce nationalist and xenophobic sentiments particularly among the economically disadvantaged. Reflecting over the 30-year period between 1914 and 1944, Karl Polanyi ([1944] 1957) interpreted the intensity of international and

---

[2] Some would say *greed* (cf. Sanders 2015).
[3] The rationality of "throw-away society" is succinctly expressed in the notion of *planned obsolescence* (Jackson 2009: 184).

ethnic conflict in these decades as reflecting a reaction to the expansion of the world market in the nineteenth century. This could also be expressed as a widespread urge to mobilize social identities that seemed to offer more collective substance than market niches, and that appeared more impervious to the dissolvent logic of the disembedded market. Ethnic conflict, nationalism, and fascism can thus be viewed as reactions to cosmopolitan modernity. Such reactions to globalization are not difficult to identify also in our own time. Given the constraints defined by the logic of money, the political left is faced with the difficult choice between siding with the neoliberals in support of globalization or with the nationalists in advocating protectionism. A third and more promising option would be to envisage ways of redesigning money to transcend this impasse by discouraging globalization and strengthening the economic conditions for building localized, place-based identities. If the degrowth movement (D'Alisa et al. 2015) is to represent a step beyond the traditional left, it will have to reconsider the implications of the money artifact that currently seems to constrain our options.

Bruno Latour (2016) has suggested that the era of globalization is over. Brexit and the American presidential election in 2016 indicated to him that the majority of people in the very countries that once launched the global market now wanted to withdraw from it and resurrect their nations. But Latour asserts that the utopia of the Globe is as unrealistic as the utopia of the Nation. He suggests that the situation would not have been much better if Hillary Clinton had won the election. However, he is finally unable to offer a third option alongside the Nation and the Globe. Does this bring us back to 1914? Are we once again forced, by the imperatives of money, to choose between the nation and the world?

The relation between the two utopias – the Nation and the Globe – is complex. In the days of the British Empire, the two were ideologically fused in concepts such as the British Commonwealth and the sentiments expressed by Britons collectively singing "Rule Britannia." British patriotism was strongly connected to colonialism. During the latter half of the twentieth century, however, there emerged a widespread understanding of world trade as emancipatory for all the participants, regardless of their economic position. Most conspicuously since the inauguration of neoliberal policies in the 1980s, free international trade has been represented as antithetical to nationalism, while the latter has been identified with regressive inclinations such as xenophobia and economic protectionism. The ideological bottom-line of this shift has been to redefine imperialism as globalization. To open world trade to increasingly asymmetric (but unacknowledged) resource flows benefitting capital accumulation in core nations is thus currently represented as global emancipation. In terms of net transfers of resources, neoliberalism can be viewed as a convoluted argument for the aggrandizement of core nations through neocolonialism. It has become more evident than ever, however, that this means the aggrandizement of their elites, leaving most of the national population far behind (cf. Bauman 1998; Goodhart 2017).

Globalization is thus ideologically represented as opposed to national-ism, even though the interests of national elites remain paramount. Hillary Clinton represents the progressive middle-class consensus that equates globalization with global solidarity. Her views are shared by a clear majority of politicians, journalists, and academics. Former Swedish Prime Minister Carl Bildt (2016) thus wrote in *The Washington Post* that Donald Trump's victory signifies "the end of the West." A similar despair recurs among most politicians and other public commentators who have been granted media space since the American election in 2016. In its eagerness to distance itself from right-wing populists, even much of the left has joined the moralizing liberal defense of globalization. But in this context, the globalist stance seems remarkably naïve. As people on the left are prone to point out, globalization generates increasing inequalities both between and within nations, increasing financial instability and vulnerability, and increasingly worrying environmental and climate change. Since 2015, 1 percent of the world's population owns more wealth than the remaining 99 percent. Meanwhile, global warming, ocean acidification, and biodiversity loss have been proceeding at record speed. This is certainly not to suggest that Trump's policies would alleviate any of these problems, but the many and widespread doubts about globaliza-tion cannot be dismissed as ignorance. Given the current trends toward economic polarization, devastating financial crises, and global warming, the neoliberal celebration of free trade is not persuasive. For several decades of globalization, inequalities, debts, and emissions have soared. The very foundations of modern society – money and fossil fuels – have revealed themselves to be bubbles about to burst.

The Biophysical Repercussions of Economic Growth

The dilemma of being compelled to choose between impoverishing the planet, on the one hand, and barricading us behind borders, on the other, is generated by the conceptual prison house of money. It is notable, however, that the most elaborate critics of growth tend to have little to say about money. Herman Daly (1996: 38, 180–185) follows Karl Marx and Frederick Soddy in exposing the illusions of money fetishism but does not appear to view the design of money as something about which we have a choice. The recipes for a sustainable nongrowth society offered by Daly (1996), Hamilton (2003), Victor (2008), Jackson (2009), Latouche (2009), Raworth (2017), and Kallis (2018) are all based on extensive modifications of the worldview of economics, but none suggest a modification of the money artifact on which the discipline is based. They represent a significant consensus regarding some fundamental criticisms of standard economics, agreeing, for instance, (1) that there are biophysical limits to growth, (2) that the North's affluence cannot be universalized, but is largely achieved at the expense of the South, (3) that there is a threshold beyond which growth is irrational and not correlated with happiness, and (4) that an absolute decoupling of growth from inputs of matter and energy is an illusion. Their solutions for building a postgrowth society, however, do not radically confront the logic defined by the artifact of general purpose money. This is not difficult to understand, given how fundamental such money is to our way of thinking and organizing society, yet nothing short of redesigning money will remedy the havoc it is playing with human society, the biosphere, and the climate.

Let us remind ourselves of some of this havoc. Johan Rockström and his colleagues (Rockström et al. 2009) have identified nine "planetary boundaries" that define the quantitative thresholds of environmental change that should not be transgressed, if we are to avoid disaster. According to these authors, three of the nine boundaries (biodiversity loss, biogeochemical flows, and climate change) had already been passed in 2009. There is currently little to suggest that the deleterious socio-ecological processes identified by Rockström and colleagues will be curbed. According to the World Wide Fund for Nature's (WWF) *Living Planet Report 2016*, a fourth planetary boundary – land-system change – had been transgressed in 2016. It also reports that vertebrate species populations have decreased by 58 percent between 1970 and 2012, and that, by 2012, the demands of human society had overshot the Earth's sustainable biocapacity by 0.6 planets. The solution advocated by WWF is "an approach that decouples human and economic development from environmental degradation," yet the likelihood of such a scenario is virtually null (Victor and Jackson 2015). The *Assessment Report for the UNEP International Resource Panel* (Schandl et al. 2016: 14–16) confirms that the vision of decoupling is contradicted by current trends. The report also highlights the severe global inequalities, observing that "the richest countries consume on average 10 times as many materials [per capita] as the poorest countries," and predicts that by 2050, nine billion people will require almost *three times* the volumes of materials consumed today, adding that such rising material use will imply the further transgression of even more planetary boundaries – and the understatement that this will impact negatively on "human health and quality of life" (ibid., 16–17). If indeed we are in desperate need of decoupling growth from resource consumption, but all evidence indicates that such decoupling is unfeasible, should we not conclude that it is growth that we must abandon?

Increasingly alarming reports on the *State of the World* are published every year. In the 2015 edition, Nathan John Hagens (2015) shows that the currently burgeoning extraction of low-EROEI fossil fuels in the United States is feasible only as long as there can be a continued expansion of financial credit.[4] If we acknowledge that the combustion of fossil fuels contributes significantly to climate change, the inevitable conclusion is that the availability of money, largely created by banks out of thin air, propels our transgression of planetary boundaries. In other words, we are allowing financial fantasies to destroy the planet. But is this at all a matter of concern for economists? Should the logic of money be conceived as a *reflection* of the material economy, or instead as its very *source*? These are the kinds of questions raised by the Anthropocene. Given the severity of the global concerns regarding planetary boundaries, current and projected ecological overshoot, escalating economic polarization, declining net energy, and financial instability, it is truly remarkable how impervious the profession of economics has remained to any external proposal that might disturb its complacent preoccupation with mathematical models that represent the suicidal trajectory of human society as predictable and on the whole reassuring. The methodological rigor[5] of those models must apparently be kept insulated from the agonies of starving people, disappearing species, and a deteriorating planet. To invert the famous phrase from Shakespeare, there clearly is madness in the method.[6]

---

[4] This expansion has been exponential since the final abandonment of a gold standard in 1971.

[5] Paul Ehrlich has used the brilliant expression "crackpot rigor."

[6] The affinity between economic logic and madness (cf. Harvey 2018) occasionally becomes particularly clear. When Lawrence Summers in his World Bank memorandum in 1991 demonstrated that "the economic logic behind dumping a load of toxic waste in the lowest-wage country is impeccable," Brazil's Secretary of the Environment responded that Summers's reasoning was "perfectly logical but totally insane" (Harvey 1996: 366–357). As Harvey concludes, the memo raises questions about "a whole mode of discourse about environmental and economic issues."

Rather than proposing to solve problems by phrasing socio-ecological conditions in terms of the obfuscating terminology of economics – for instance, employing hybrid terms such as "natural capital" and "ecosystem services" (cf. the 2005 Millennium Ecosystem Assessment report *Living beyond Our Means*) – or by following the World Economic Forum in identifying the main global "risks"[7] of highest concern (cf. World Economic Forum 2014), it is incumbent on us to focus on that peculiar artifact that *generates* the terminology and outlook of economics. That artifact – general purpose money – is implicit but unassailable in any standard textbook on economics. However, I repeat that it is not an incontrovertible product of nature, but a recent social construction, that can in principle be redesigned.[8] It is widely acknowledged that the operation of general purpose money and its intrinsic imperative of economic growth (some would say "capitalism") tends to generate increasing global inequalities (Piketty 2014; Alvaredo et al. 2017). Even if it is politically unfeasible to consciously pursue degrowth as a social strategy, the apparent inevitability of decline could be harnessed to the goals of sustainability, resilience, and global justice. This presupposes concerted preparations for the plausible occurrence of economic decline, including training in skills likely to be in high demand in communities resorting to greater self-sufficiency, as well as public information on the long-term benefits of degrowth for sustainability, justice, and the vitalization of resilient human communities. Most centrally, we need more awareness among economists and laypeople alike about the derivation of the growth compulsion from the specific properties of the artifact of general purpose money, and about the fact that those specific properties are as potentially mutable as the rules of a board game.

---

[7] Appropriate questions to the World Economic Forum are: "Risks" to *what*? If affluence, *whose* affluence? This is not to say that the World Economic Forum was not correct in identifying a major financial collapse as the most troubling global risk in 2014.

[8] But then, of course, most economic models would have to be redrawn and most economists retrained. Perhaps not an unreasonable price, after all, for a sustainable planet?

6

The Ontology of Technology

A central theoretical challenge that has haunted me ever since the 1970s is how to articulate and pursue the conviction that our most tangible manifestation of capital accumulation – modern technology – is to be understood as a global phenomenon (Hornborg 1992, 2001). Many people would immediately agree that such an observation is historically valid, but my ambition has been to give it a more literal significance than reflected in conventional acknowledgments that the emergence of industrial technology was conditioned by global processes (e.g., Wolf 1982; Marks [2002] 2015). My reason for devoting this chapter to the ontology of technology is my conviction that mainstream modern perceptions of what technologies basically *are* derive from the distorted understanding of technological progress that emerged from the historical experience of nineteenth-century Britons in the core of their colonial empire. The ontological shift required to grasp the scope of this distortion demands the readiness to deconstruct familiar categories that is the hallmark of the "ontological turn" in anthropology, but paradoxically also the kind of analytical reason that is associated with the very Enlightenment that the ontological turn repudiates. As I shall show in the following chapters, the illusions of machine fetishism continue to constrain our deliberations on the potential of our species to avert environmental disaster, for instance by shifting to renewable energy (Chapter 7) or to a socialist economy in which technologies are meant to serve the interests of justice and sustainability (Chapters 8 and 9). My intention, in advocating a fundamentally revised ontology of technology, is to expose the hegemonic illusions of machine fetishism.

The extent to which the industrialization of British textile production was contingent on global relations of exchange has been illuminated in detail by historians such as Joseph Inikori (2002) and Sven Beckert (2014). However, no matter how willing we are to concede that access to advanced technology represents a global, historical privilege, it will require a fundamental ontological shift to conceive of modern technology *as such* as a global societal phenomenon. To do so means acknowledging not only that the social is a product of the interpenetration of the local and the global, which is the hallmark of global perspectives such as world-systems analysis, but also that technology is a product of the interpenetration of the material and the social. While the local/global interface has been explored and reconceptualized within world-system perspectives, the material/social interface implicates a range of approaches within the sociology, history, and philosophy of technology that have emerged out of the concept of "sociotechnical systems" (Bijker et al. 1987).

These latter approaches were largely established in the 1980s and 1990s, and they contributed to a renewed anthropological interest in technology and material culture. This was a period of significant progress in theorizing the social dimensions of technical artifacts, but I shall argue that the commitment to a case study approach has precluded a global understanding of the phenomenon of modern technology. Conversely, to reconceptualize technology by transcending the distinction between the material and the social poses a significant challenge for a global social science. I shall begin by reiterating the outlines of my argument, before more systematically discussing the insights and limitations of some foundational attempts to rethink the relation between technology and society.

Technologies as Material Manifestations of Global Society

Methodologically, the reconceptualization of industrial technology as a global phenomenon is difficult to accomplish through conventional anthropological fieldwork or even archival studies. The global processes that generate modern technologies cannot be exhaustively accounted for through local perspectives, regardless of where empirical investigations are conducted. Even the comparatively simple case of the eighteenth-century British steam engine implicates historical world market prices of a number of commodities, the ratios at which several biophysical resources (including embodied labor) are exchanged, the balance sheets of various market actors, and so on.[1] Inasmuch as modern technological systems diffusely implicate myriad market transactions throughout the world, they are interfused with the global social fabric and simply not amenable to the case study approach. More fundamentally, a global, metabolic rethinking of technology requires a theoretical analysis that transcends not only historical, sociological, and ethnographical narratives but also mainstream perspectives in engineering, economics, economic history, sociology, and other social sciences. More than meticulous empirical demonstration, the perspective I am proposing requires a gestalt shift in perception. Across the social sciences, the commitment to empiricism and the case study approach precludes a genuinely global perspective on the ontology of technology. If abstraction and generalization are as crucial to theoretical progress in the social as in the natural sciences, and if the object of study in the social sciences is recognized as global in scope, the case study approach may not be a sufficient foundation for the theorization of global sociometabolic processes.[2]

It is incumbent on me, at this point, to more precisely define what I mean by "technology." In this book, I restrict the definition to *a system of material artifacts that locally augments the capacity of a certain social category in some respect (e.g., by reducing necessary expenditures of human time), while being contingent on the rates at which biophysical resources (including embodied labor) are exchanged on the world market.*

---

[1] Joseph Inikori (1989: 367–369) has demonstrated that "the conditions for the technological innovations in the [cotton textile] industry in the late eighteenth century" were created by "the production of cotton textiles for the slave trade in Africa and for the clothing of the African slaves on the New World plantations." This observation provides a welcome counternarrative to Eurocentric historiographies of technological progress, but – unless combined with theoretical insights on "sociotechnical systems" (cf. Hughes 1987; Pfaffenberger 1992) – does not in itself transform our ontological framing of technology. Even explicitly non-Eurocentric historiographies of technology (e.g., Blaut 2000) tend to focus on the non-European *origins* of particular innovations, rather than on technologies as components of systems of social reproduction that transcend Europe.

[2] There is a vast literature documenting in rich detail how specific technologies organize social relations and perceptions in various cultural and historical contexts. Such studies are indispensable, but they illustrate our propensity to prioritize empiricist knowledge production. The fact that a massive historical or ethnographic treatise might be succinctly summarized in a single theoretical paragraph indicate that different forms of knowledge should be understood in terms of levels of abstraction or "logical types." The potential volume of evidence for a particular phenomenon chosen as the object of research may be almost infinite, and we tend to be more impressed the more elaborately detailed it is, yet the pursuit of empirical detail encourages a continuous proliferation and fragmentation of research, precluding the kind of transdisciplinary, theoretical syntheses that would serve humanity well in its current predicament.

It should be observed that most preindustrial artifacts do not meet the latter criterion. However, from eighteenth-century British imports of Swedish iron and American cotton to contemporary Euro-American imports of Chinese electronics and Brazilian ethanol, the operation of modern technology is ubiquitously contingent on asymmetric exchanges of embodied labor and land, veiled by the fictive reciprocity of market prices. The discontinuity in material culture between preindustrial tools, on the one hand, and industrial machines contingent on global market prices, on the other, appears to coincide with the emergence of a completely new form of technological rationality that has intrigued generations of social thinkers from Karl Marx through Jacques Ellul to Langdon Winner. It defines the "rupture" between premodern and modern technology that was identified by these thinkers. This modern rationality, premised on a virtually unrestricted scope for outsourcing or *displacement* of resource requirements, is the local manifestation of the new logic of technology unleashed by the global economy since the late eighteenth century. Both as social practices and as concepts, technology and economy have emerged in tandem, recursively extending the scope for mobilizing human and natural resources to aggrandize the power of a global minority.

Paradoxically, our firmly entrenched inclination to insulate the material from the globally constituted social can be illustrated by none other than Karl Marx's perception of early industrial machinery.[3] In his deliberations on machinery, Marx clearly approaches it as a *local* consideration:

The use of machinery for the exclusive purpose of cheapening the product is limited by the requirement that less labour must be expended in producing the machinery than is displaced by the employment of that machinery. (Marx [1867] 1976: 515)

Although he is abundantly aware that labor and machinery are transposable in the sense that labor both produces and is displaced by machinery, and that this convertibility is made possible by money, Marx does not address the extent to which the very existence of machinery is contingent

---

[3] This is paradoxical because Marx pioneered the theoretical integration of the material and the social, as in his understanding of machinery as "congealed labor," but did not sufficiently acknowledge the global societal context of technological development. His focus was on production, but he rarely asked for whom (i.e., for which markets) production occurred, or from where derived the labor and land embodied in raw materials such as cotton fiber. In this respect, Inikori's (2002) research on the globalized incentives for – and prerequisites of – British mechanization is an important corrective to the myopic (i.e., theoretically *incomplete*) Marxist concept of "productive forces."

on global exchange relations. The assumption is that both types of labor – the labor expended in production of machinery and the labor displaced by it – have the same cost. The relative cost of labor in different geographical areas or nations is not discussed as a factor influencing the adoption – not to mention the existence – of industrial machinery.[4] Nor is the accumulation of machinery understood as contingent on flows of natural resources. Yet we know, for instance, that the annual British import of Swedish iron in the 1760s represented around a million hectares of Swedish forest and almost 14,000 person-year equivalents of Swedish labor (Warlenius 2011: 68). The relative cost of British and Swedish land and labor should thus have been a crucial consideration determining the extent of British investments in machinery in the 1760s. Such considerations of global price differences, and the asymmetric exchanges of embodied labor and land that they made possible, must have been fundamental not only to the feasibility of early industrialization in England but also have continued to determine the conditions for technological development for two and a half centuries.

However, nowhere in the voluminous literature on the history, philosophy, and sociology of technology have I found an indication of the profound implications that this should have for the very ontology of modern technology.[5] The closest thing to such a paradigm shift that I have encountered is Andre Gunder Frank's (1998: 204) insight that "technological development was a *world economic process*, which took

---

[4] Neither has Emmanuel's (1972) analysis of the unequal exchange of embodied labor in world trade contaminated the Marxist perception of technology. To enhance the productivity of local labor by means of technologies that are contingent on net transfers of resources and embodied labor from elsewhere thus continues to be viewed by Marxists as a fundamentally beneficial process leading ultimately to a postcapitalist society (see Chapter 9).

[5] A representative definition of technology by a centrally positioned authority on its ontology is Wiebe Bijker's (2010: 64) assertion that it "comprises, first, artefacts and technical systems, second the knowledge about these and, third, the practices of handling these artefacts and systems." How, one wonders, do exchange relations on the global market – distributing the material resources that are requisite for the technology to *exist* – at all enter the picture? Michel Callon (1987: 90, 94) briefly touches on the issue when he mentions the market prices of oil and metals as included among the various "forces" or "actors" that influenced the competition between rival automobile designs in France in the 1970s, but this acknowledgment of market factors does not prompt him to rethink technology as fundamentally a matter of distribution. Callon's (1998) edited volume on the sociology of markets – which, like Latour, frequently evokes anthropology – similarly deals with the conditions of "calculative agency," rather than the asymmetric material flows generated by discrepancies between the substance and the price of commodities. Given the professed concern of Actor-Network studies with the material outcome of processes involving the interaction of human and nonhuman entities, this is a significant omission.

place in and because of the structure of the world economy/system itself"
(emphasis in original). Such insights nevertheless tend to be compatible
with an intact demarcation of the materiality of machines vis-à-vis the
relationality of global exchange.[6] The demarcation of the technological
and the social is strengthened by the conventional conviction that technology is fundamentally a matter of *knowledge* of the physical properties
of nature.[7] But, as I have argued elsewhere (Hornborg 2016: 9–16),
although engineering knowledge is a necessary condition for technology,
it is not a *sufficient* one. Much as a biological genotype is not synonymous
with an organism, a technical blueprint remains an abstraction until it is
embodied in a metabolic process.[8] The feasibility of the metabolic flows
that sustain an organism is determined by its ecological context, whereas
the feasibility of the metabolic flows that sustain a modern technology is
determined by the world market. The reification and sequestration of the
eighteenth-century British steam engine as an autonomous material artifact, contingent merely on the ingenuity of its inventors, was an illusion –
as our understanding of technology continues to be to this day. Its
ultimate foundation was the social organization of the eighteenth-century
world-system. Our inclination to objectify technology as detached (or
excised, as it were) from its sociometabolic context is what I have referred
to as "machine fetishism" (Hornborg 1992). In reinserting the steam-
driven textile factories of nineteenth-century England into the asymmetric
metabolic flows of the triangular Atlantic trade, we can reinterpret them
as social instruments for globally appropriating embodied human labor
and natural space.[9]

---

[6] Although several other authors (e.g., Wilkinson 1973; Pomeranz 2000) have acknowledged that the Industrial Revolution entailed a British displacement of resource demands
overseas, such insights do not seem to have contaminated their fundamental conception
of technology as a politically neutral, extrasocietal phenomenon representing the convergence of physical properties of nature and human skill in harnessing such properties.

[7] Cf. Layton 1974; Mitcham 1994: 192–208. Ingold (1997: 130) captures this modern
conception of technology as "the means by which a rational understanding of [the]
external world [of nature] is turned to account for the benefit of society."

[8] Cf. Ingold 1997: 112.

[9] The traditional ontological sequestration of the natural from the social seems to have
constrained Marxist theory from grasping the *socionatural* constitution of machinery.
Rather than politically neutral revelations of nature, the first technologies for harnessing
fossil energy were components of global strategies for displacing work and environmental
loads onto remote cotton plantations. The requisite capital to invest in fossil-fueled technologies to this day represents indirect land and labor requirements beyond the horizons of the
investors.

The Materiality of Technology as Criterion of Its "Naturalness"

My use of the concept "sociometabolic" highlights, again, that there are at least two conceptual demarcations that we need to transcend: not only the local versus the global as jointly constitutive of the social but also the material versus the social as jointly constitutive of technology. Its materiality has prompted us to classify technology as revealed nature. As Bruno Latour, Tim Ingold, and others have argued, modern worldviews have suffered from a misleading dualism that opposes society and nature as ontologically separate domains. Such dualism has constrained our capacity to recognize the mutual interpenetration of the social and the natural in tangible physical phenomena such as landscapes, bodies, and technologies.

However, the argument for an *analytical* distinction between nature and society can be illustrated precisely by landscapes, bodies, and technologies. Environmental historians know that domesticated landscapes everywhere reflect a social order: the layout of fields, buildings, roads, and so forth mirror the organization of society. Anthropologists are similarly aware that human bodies everywhere reflect their positions in society: their occupation, diet, health, and so on. In neither case would it be reasonable to deny that there is also a natural substratum, whether ecological or genetic. As we turn to the third illustration – technologies – we arrive at the central problem in rethinking the relation between society and nature. In utilizing physical regularities such as the principles of thermodynamics, technologies obviously have natural aspects, but what exactly do we include in the common observation that they are also social phenomena? What does it mean to say that technologies are socially constructed (cf. Bijker et al. 1987)?

It is obvious that the designs of technologies not only *reflect* their social contexts, performing specific functions that are in demand in those contexts, but conversely also mediate and *organize* social relations. Technologies may locally replace, reorganize, and control labor; increase rates of profit; serve as tools of military conquest; and generally shape human mentalities and perceptions of reality. Such observations permeate studies of the history and sociology of technology, whether conducted within the traditions of SCOT (The Social Construction of Technology; see Bijker, Hughes, and Pinch 1987), STS (Science and Technology Studies, e.g., Latour 1987), Marxism, phenomenology, or other contributions to the philosophy of technology. But none of these approaches acknowledges the extent to which technologies may presuppose and embody specific patterns of social organization at the *global* level. They tend to focus on the specific designs and local human consequences of given technological systems, but do not theorize their global societal conditions in terms of the requisite ratios of resource exchange on the world market. Yet it is incontrovertible that a given machine utilized in production represents a net input of physical resources, and that its output of commodities on the market must fetch exchange-values greater than those of the inputs of resources required to keep it running.

Building on Nicholas Georgescu-Roegen's (1971) insight that economic processes that enhance utility simultaneously increase entropy, I have inferred that exchange-values and productive potential must be inversely correlated, and that the uneven accumulation of technological infrastructure signifies the confluence of thermodynamics and imperialism (Hornborg 1992). It would be unreasonable to deny that the principles of thermodynamics derive from nature, while the strategies of imperialism derive from global society. Clearly, however, mainstream economists are convinced that their accounts of growth and technological progress have no use for thermodynamics. In their worldview, nature is irrelevant for the constitution of society. No less myopic, however, is the inclination of mainstream engineers to believe that the accomplishments of technology are independent of the exchange rates organizing global resource flows. In their worldview, technology is tantamount to the revelation of nature and apparently not contingent on the organization of society.

The Steam Engine as Embodiment of Global Relations

Whereas earlier efforts to theorize the social contexts and consequences of technology tend to relate a specific technological artifact to a restricted range of immediate users, the approach advocated here is to view its very existence as contingent on global social relations of exchange. I return to the example of the steam engine employed in British textile factories in the early nineteenth century. It owed its existence not only to the ingenuity of inventors like James Watt but fundamentally also to the lucrative structures of exchange between Britain, West Africa, and America. Without the trans-Atlantic flows of embodied African labor and embodied American land, and the African and American markets for British textiles, it is difficult to imagine a British Industrial Revolution (cf. Inikori 2002).

In the category "social," as previously declared, I include the organization of the world-system, rather than restricting the concept of society to a national or local phenomenon. At this global level, technologies are indeed infused with social organization in a sense that tends to be insufficiently acknowledged in conventional thought. Whereas other approaches to the social dimension of technology might focus, for instance, on the local social negotiations leading to the optimally designed engine, the factory owners' use of steam engines to discipline labor, the use of steam engines as tools of British imperialism, or workers' experiences of labor in steam-driven factories, none of these approaches problematizes the ontology of technology. They assume the existence of an artifact like the steam engine as given by the strategies of engineers, capitalists, and other human actors, reflecting their purposes and imposing their designs on other people, but they do not reflect on the steam engine as an embodiment of the world-system that made it possible to begin with. Nevertheless, as I argued previously, much as a living organism is a literal embodiment of the biophysical flows of energy and matter that sustain it, a functioning machine is an embodiment of biophysical resource flows, orchestrated by money and prices.

Although several philosophers have emphasized that some material phenomena, such as living organisms, should be understood in *relational* terms – as manifestations of wider webs of relationships – this insight remains to be extended to the material existence of technological artifacts. Perspectives on technology such as those of Latour (1987) and Ingold (2000) are often referred to as relational, but neither of these authors considers the concrete metabolic flows and asymmetric material resource transfers through which a modern technological system is reproduced. As the organization of such flows and transfers is orchestrated by means of market prices, this requires a consideration of political economy. David Harvey (1996: 257) observes that relational perspectives explicitly challenge "the famous mechanistic theory of nature with which science had broadly remained content since the seventeenth century, mainly because it worked so well across a broad terrain of technological practices." Although the machine has long served as a model of nature, the argument here implies that it cannot even *itself* be exhaustively accounted for with a mechanistic narrative. This insight constitutes a fundamental challenge to the mechanistic worldview.

The identification of an industrial technology such as steam power with the asymmetric metabolic flows of the global social system that made it possible prompts us to rethink conventional narratives of industrialization and technological progress. Although locally perceived as morally and politically innocent progress, from a global perspective the Industrial Revolution assumes the appearance of an asymmetric exchange of

COMMODITY	VOLUME FOR £1000 IN 1850	LABOR TIME EMBODIED	LAND AREA EMBODIED
Raw cotton	11.84 tons	20,874 h	58.6 ha
Cotton cloth	3.41 tons	14,233 h	0.5 ha

FIGURE 4 The asymmetric exchange of human time and natural space obscured by world market prices of raw cotton and cotton cloth in 1850

embodied labor time and a massive environmental load displacement.[10] I have referred to such global processes underlying the Industrial Revolution as "time-space appropriation" (Hornborg 2006, 2013a; figure 4). My point has been that the phenomenon of modern technology to a large extent can be conceptualized as a local saving of time and space at the expense of human time and natural space lost in other parts of the world-system.[11]

A common objection to this line of reasoning is that the technology "as such" does not necessarily imply inequitable exchange relations or environmental load displacement. This was certainly the view held by Karl Marx, who predicted that the productive forces generated by capitalism could be extricated from their capitalist context and employed for collective benefit under socialism. He apparently perceived technology as innocent in itself. Given more recent social-science insights on the inextricable connections between material artifacts and their social contexts, however, such a detachment of a technology from the exchange relations

---

[10] In 1850, a sale of British cotton manufactures on the world market for £1,000, and the purchase of raw cotton for the same amount, would have been tantamount to exchanging 14,233 hours of embodied British labor for 20,874 hours of Afro-American labor, and of less than a hectare of embodied British land for 58.6 hectares of American land (Hornborg 2013a).

[11] In terms of "the black box" evoked by Pinch and Bijker (1987) and Winner (1993: 365), a modern technology in this view is a device for converting inputs of human time and natural space deriving from one social category into outputs in the form of power (e.g., a liberation of time and space) for another.

that spawned it is implausible. To imagine the steam engine as detachable from asymmetric global resource flows indeed deserves to be referred to as "machine fetishism." Extending Marx's observations on money and commodity fetishism to the field of technology, I have proposed that machines, like money and commodities, are global social relations of exchange masquerading as things. To metaphorically summarize my argument so far, the globalized ensemble of technological artifacts is the physical scaffolding (the morphology, so to speak) of world society, while the flows of money and resources are its physiology. The functions of engineering science and mainstream economics are to keep this global metabolism operating efficiently. In both respects, humans enlist physical materials and energy to organize their social inequalities, but in neither field are the ratios by which such physical resources are appropriated perceived as relevant to the validity and efficacy of its basic tenets. As we shall see, the implications of the observation that technology is "socially constructed" are very different depending on at which scale it is made.

Some Formative Contributions on Sociotechnical Systems

The role of technology in human societies has been a central focus of two and a half centuries of reflections on social change, beginning with the dramatic transformations of society in nineteenth-century England. Whether celebrated as progress or demonized as destructive of human values, technological development has been recognized as inextricably intertwined with the organization of society and human consciousness.[12] The ambivalent attitudes toward technology are apparent in Robert Post's (2010) reflections on 50 years of the Society for the History of Technology (SHOT) and its journal *Technology and Culture*. Although aspiring to transcend the technical interests of mainstream engineers and antiquarians, there were nevertheless deep concerns in 1959 that the word "culture" would deter engineers from subscribing. In the deliberations that finally settled on including the concept in the journal title, advocates referred to iconic anthropologists such as Margaret Mead, Ruth Benedict, and even Edward Tylor.[13] According to the historian Melvin Kranzberg,

---

[12] There have even been attempts to understand the historical development of technology as a fundamentally religious pursuit (Noble 1999).

[13] See also appeals to anthropology by the historian John Staudenmaier (1984: 708) and the "engineer-turned-sociologist" Wiebe Bijker (2010: 63). It is interesting to note that historians and engineers seem to have been more enthusiastic about an "anthropology of technology" than were anthropologists (see following text).

who founded *Technology and Culture*, its title reflects "the broad socio-logical, anthropological, economic, and humanistic content of the jour-nal" (ibid., 977). There was clearly an awareness that there is more to technology than the technical, and that it should not simply be left to engineers.

In retrospect, the new society and journal were compelled to strike a balance between criticism and public acceptance. Early contributors to the journal included wide-ranging, critical, and often pessimistic writers like Lewis Mumford, Jacques Ellul, Lynn White Jr., Walter Ong, Kenneth Boulding, Thomas Hughes, Joseph Needham, Aldous Huxley, and Wernher von Braun. It seems that the kinds of humanistic and ecological criticism of technology delivered by these people – such as in the special issue of *Technology and Culture* devoted to the records of the Encyclopedia Britannica Conference on the Technological Order (Stover 1962) – were finally not subversive enough to be incompatible with Kranzberg's success in "enlisting support from a wide array of academic notables" (Post 2010: 979) including Thomas Kuhn, David Riesman, Robert Merton, and the influential management theorist Peter Drucker.[14] In successfully seeking allies "from beyond academe in engineering and industry" (ibid.), *Technology and Culture* could obviously not be seriously subversive. Its vision was to reveal the socio-cultural contexts of technological development, but only with regard to humanistic concerns or the designs of specific technologies, never in the sense of identifying the asymmetric global metabolic flows on which modern technology *as a whole* is contingent. It is an ironic illustration of the very sequestration of technology from society that these and subsequent critics have pervasively addressed that "technology" to this day remains a legitimate target of critique, as long as the criticism is not extended to the social organization (the capitalist world market) that it embodies.[15]

---

[14] Drucker in 1967 was appointed as SHOT representative to the American Association for the Advancement of Science (AAAS) (Post 2010: 990). Clearly, the discourse promoted by an organization represented at the AAAS could not possibly challenge technological progress as an index of inherently exploitative capital accumulation.

[15] Reflecting on 25 years of SHOT and *Technology and Culture*, Staudenmeier (1984: 722) notes "the complete absence of a Marxist perspective" and the striking omission of discussions on "the influence of capitalism on technology." This is no doubt related to his finding that, of 272 articles published in the first two decades of the journal's existence, only *two* consider "non-Western technologies from a contextual perspective" (ibid.). Twenty-five years later, his critical assessment is echoed by Wiebe Bijker (2009).

More than 40 years ago, Langdon Winner (1977) reviewed some core concerns of philosophers of technology such as Marx, Heidegger, Mumford, Ellul, and Marcuse. For two centuries, a pervasive theme among philosophers and people in general has been the apprehension that technological change is a process beyond social control, pursuing its own autonomous logic rather than answering to the needs and wishes of humans. Winner traces recurrent worries that humans are becoming the servants of machines, rather than their masters. Common to these humanistic concerns is the observation that technical devices tend to have significant and often unanticipated consequences for social relations among the people who use them.[16] In a later book explicitly critical of the political aspects of technology, Winner (1986) begins by observing that human lives are shaped by an interconnected technological order that "transcends national boundaries," yet generally illustrates his "political philosophy of technology" with American examples.[17] Like his predecessors from Marx to Marcuse, Winner is concerned with how technologies shape the lives, minds, and societies of those who have access to them, rather than with how they shape *global* relations between those who do and those who don't. However, his focus on how material culture embodies and reproduces social relations of power and inequality can be extended to include such global considerations.[18] Winner's "epistemological Luddism" (1977: 330) is not reducible to "blaming the hardware" (1986: 20) but recognizes that sociotechnical systems frequently have political implications.

[16] Whereas general optimism about technological progress has been explicit and straightforward since the eighteenth century, the pessimism voiced by skeptics like Mumford, Ellul, and Marcuse tends to be more impalpable, suggesting intangible intuitions about a fundamental moral deficiency underlying the modern technical domination of the world. Their humanistic critique of modern technology is often dismissed as romanticism, but it potentially converges with a rethinking of sociotechnical systems as obscure strategies for redistributing human time and natural space in world society. The doubts about technical artifice may thus reflect not only humanistic and aesthetic concerns but also moral intuitions about the insidious distributive politics that it embodies.

[17] In the chapter titled "Do Artifacts Have Politics?" Winner (1986: 22–23 [reprinted from Winner 1980]) shows that the low bridges over parkways on Long Island, New York, were deliberately designed to exclude buses (and their lower-class passengers) from a public park.

[18] The dependence on global exchange rates can be seen as a third way in which technologies are "inherently political," in addition to Winner's (1986: 32) distinction between how a technical system may *require* a particular social environment, on the one hand, and how it may be *conducive* to it, on the other.

A foundational volume edited by Wiebe Bijker, Thomas Hughes, and Trevor Pinch (1987) gathered several detailed case studies of the sociological processes that have shaped particular technologies. In the same year, Bruno Latour (1987) published a volume that was to become paradigmatic for the methodological approach of STS. Both books represent a shift of focus from the traditional concern with the social consequences of technology to how technologies are shaped by society. The chapter by Pinch and Bijker (1987) in the former volume offers an influential sociological methodology for studying SCOT. Hughes's (1987) chapter introduces the concept of "sociotechnical systems," which analytically integrates the social and the technological.[19] Michel Callon's (1987) chapter outlines the basic tenets of Actor-Network Theory, an approach closely aligned – and subsequently identified – with that of Latour, which controversially claims that technical artifacts and other inanimate factors have agency in much the same way that humans and other living beings do. Also pioneering the Actor-Network Theory approach, John Law's (1987) chapter discusses the role of naval technology in Portuguese maritime expansion in the fifteenth and sixteenth centuries.[20]

In this same year, Latour and primatologist Shirley Strum (Strum and Latour 1987) published a co-authored paper that argued that the reason why humans can build more permanent and extensive societies than baboons is that humans can stabilize their social relations by anchoring them to artifacts of various kinds. This perspective is fundamental to Latour's later work on the role of technical artifacts as mediators of human social relations.

---

[19] For a retrospective appraisal of the consolidation of SCOT and the significance of the concept of the "sociotechnical," see Bijker (2010).

[20] Although implicating a world-system perspective, Law's (1987) chapter deals with the Portuguese navy as a tool of conquest, rather than as a form of capital accumulated by means of economic expansion. The ambition is to provide a holistic view of the sociotechnical "network" enabling Portuguese expansion, but Law's approach is ultimately constrained by his lack of attention to the global metabolic requisites of capital accumulation (here in the form of ocean-going ships and their artillery). Relevant questions would be from where and at which exchange rates Portugal was able to obtain the timber for its caravels and the bronze for its cannons, and how this reflected the relative prices of labor and land in the sixteenth-century world. The same omission recurs in Headrick's (1981, 2010) discussions of subsequent military technologies as "tools of empire." Given our inclination to distinguish the material and the social, even aspirations to merge these aspects tend to restrict themselves to how the *designs* of technical artifacts are shaped by incentives generated by the societal context, ignoring how the very existence of a technology is contingent on specific sociometabolic flows.

An Anthropology of Technology

This is also the time when anthropologists, after decades of silence, begin to express a renewed interest in technology and material culture. In "an excursion in the philosophy of technology," Tim Ingold (1988), drawing on Marx, reflects on how the transition from skill- and tool-based manufacture to machines and dehumanizing "machinofacture" involved "a progressive *objectification* and *externalization* of the productive forces" (emphasis in original). The following year, in a paper originally published in 1990, Ingold unravels and problematizes how "technical relations have become progressively disembedded from social relations, leading eventually to the modern institutional separation of technology and society" (Ingold 2000: 321–322).[21] He writes:

> Having thus been placed outside of society and culture, technology could – so far as most anthropologists were concerned – be safely ignored. It was considered to be just one of those things, like climate or ecology, that may or may not be a determining factor in human affairs, but whose study can be safely left to others. As climate is for meteorologists and ecology for ecologists, so technology is for engineers. (Ibid., 313)

At this time, also, Bryan Pfaffenberger (1988a: 236) proposed an "anthropology of technology" that views technology as "a total social phenomenon in the sense used by Mauss; . . . simultaneously material, social and symbolic." He suggested that the modern concept of technology – "a term that stands, arguably, at the very centre of what Westerners (and Westernised people) tend to celebrate about themselves" – is a "mystifying force of the first order," denoting a variety of fetishism concomitant to the fetishism of commodities revealed by Marx (ibid., 237, 250). Pfaffenberger attributes to Marx the "extraordinary anthropological insight [that] *the Western ideology of objects renders invisible the social relations from which technology arises and in which any technology is vitally embedded*" (ibid., 242; emphasis in original). A recurrent theme in Pfaffenberger's papers from this period is that traditional Hindu temples in Sri Lanka can be understood as a kind of technology that is no more mystical than a modern machine

---

[21] In a later paper, Ingold (1997: 108) demonstrates how this understanding of the historical emergence of a concept of disembedded "technology" is completely parallel to anthropological analyses of the emergence of the disembedded "economy."

(Pfaffenberger 1988b).[22] Drawing on Latour[23] and Actor-Network Theory, he suggests that the Industrial Revolution was fundamentally about replacing "ritual machinations like the temple" with "kinematic machinations such as the windmill or watermill" (ibid., 15). Like other sociotechnical systems, the temple regularizes and disciplines human behavior. In a subsequent review article, Pfaffenberger (1992) advocates a renewed anthropological interest in technology and material culture building on the concept of sociotechnical systems as applied by Hughes (1987), Latour (1987), and Law (1987):

> As it stands, a topic with which anthropology was once closely identified – the cross-cultural study of technology and material culture – has been largely taken up by scholars working in other fields, such as the history of technology and the interdisciplinary field known as science and technology studies (STS). (Pfaffenberger 1992: 492)

Pfaffenberger declares that the central objective of his review article is to "convey the sociotechnical system concept to an anthropological audience," a concept that "refuses to deny the *sociality* of human technological activity" (ibid., 493; emphasis in original).

With an edited collection of case studies published the following year, Pierre Lemonnier (1993) encouraged anthropologists to approach technologies as complex sociocultural phenomena that extend far beyond the merely technical.[24] The collection includes chapters by Ingold, Pfaffenberger, and Latour. Latour's (1993b) chapter reflects on his 1988 study of the negotiations around the aborted subway project Aramis, planned to integrate southern Paris. In contrast to primate societies like the baboons

---

[22] While it is incontrovertible that traditional Hindu temples integrated complex sociotechnical systems – illustrating the premodern embeddedness of technical relations in social relations highlighted by Ingold – this is precisely *not* equivalent to saying that temples are machines (in Ingold's sense). However, Pfaffenberger's (1992: 501) dismissal of Spier's (1970: 2) distinction between technology and magic is entirely valid.

[23] Pfaffenberger (1988b: 14, 1992: 510) cites Latour's (1987: 129) apposite definition of a machine as "first of all a machination, a stratagem, a kind of cunning."

[24] Upon searching the Internet for contributions to an "anthropology of technology," I have found very little beyond those of Ingold, Pfaffenberger, and Lemonnier dating to the 1980s and 1990s. A possible explanation is that the insight that specific systems of technical artifacts must be understood as embedded in social relations and cultural frames of mind (i.e., that they should be approached as sociotechnical systems) is so obvious that it no longer needs to be asserted. Indeed, already in the early 1990s, Winner (1993: 376) felt that constructivists had been repeating their point about the social context of technological development "ad nauseam."

he studied with Shirley Strum, he concludes, human societies are stabilized by the delegation of tasks to nonhuman components:

> Techniques are not something around which there is a society. It [*sic*] is society considered in its obduracy. It is society *folded*, society made durable, society made complicated in order to resist more tensions by enrolling more non-humans ... Although this folding, this detour, this shifting down, this embedding is clear in anthropologists' accounts of exotic technologies, it is not so obvious in modern high-tech cases. (Latour 1993b: 379–380; emphasis in original)

Deploring that social scientists have "used the Durkheimian model on everything *but* science and technology," Latour argues for an extension of "social constructivism to science and technology" (ibid., 392–394; emphasis in original).[25]

Sociotechnical Systems as Power

In the same year, Langdon Winner countered that the empiricism of social constructivism is not conducive to "one's ability to talk in penetrating, reliable ways about modern technology *in general*" (Winner 1993: 363; emphasis added). Its rejection of "the abstract speculations of philosophers" such as Marx, Ellul, Heidegger, Mumford, and Illich in favor of detailed empirical case studies tends to lead to a disregard not only for the social consequences of a given technology but also for issues regarding "the broader distribution of power in society" (ibid., 364, 368). Winner criticizes social constructivism for providing "no solid, systematic standpoint or core of moral concerns from which to criticize or oppose any particular patterns of technical development," and for implicitly subscribing to the view that "what matters in the end is simply the exercise of raw power" (ibid., 374). According to Winner, the projects of social constructivists are "carefully sanitized of any critical standpoint that might contribute to substantive debates about the political and environmental dimensions of technological choice," risking a retreat into "a blasé, depoliticized scholasticism" (ibid., 375–376). A quarter of a century later, Winner's indignation rings truer than ever. His persistent observation that a given technology that provides benefits to some may cause harm to others is applicable even to the global social metabolism of modern technology as a totality.

---

[25] Like so much else (e.g., truth, knowledge, logic, theory, even society), Latour has since repudiated constructivism (see Harman 2007).

We have traced the emergence and consolidation,[26] in the 1980s and 1990s, of a discourse that scrutinizes how specific technological artifacts are imbued with social relations, symbols, ideas, intentions, even politics, but that completely ignores the question of how the aggregate phenomenon of technology operates as a sociomaterial system for organizing and reproducing power and inequalities in global society. While the anthropology of technology has generally applied perspectives similarly constrained by the case study approach, a global anthropology capable of transcending the distinction between the social and the material might recognize the planetary technomass as the physical scaffolding that buttresses or "stabilizes" the structure of world society. Ultimately, the most significant sociotechnical system is nothing less than the world-system.[27]

The contributions reviewed so far, whether by philosophers, historians, sociologists, or anthropologists, have all shared the insight that technology represents the interpenetration of the material and the social. In the restricted sense explored by social constructivism, this was established decades ago. However, in none of these four intersecting fields – the philosophy of technology, the history of technology, sociological studies of science and technology, and the anthropology of technology and material culture – have I found a concern with the global distributive dimension of the sociometabolic phenomenon of modern technologies as sociotechnical displacement strategies.[28] The question was not a concern of any of these fields 30 years ago, nor has it become a concern for any of them in the time that has elapsed since then. This conclusion is based on perusal of several texts that can be expected to reflect the diversity of topics within their respective fields, whether the philosophy of technology (Ihde 1993, 2010; Mitcham 1994; Lawson 2008; Kirchhoff 2009; Scharff and Dusek 2014), the history of technology (Smith and Marx 1994; Nye 2006; Headrick 2009), science and technology studies (Ihde and Selinger

[26] To apply the constructivists' own terminology, we might say "closure."
[27] While Mumford's slave-propelled "megamachines" of antiquity and the modern world-system are both sociotechnical systems, we are currently unable to see how technologies in the latter case are equally contingent on power inequalities.
[28] There is, of course, recurrent mention of global problems – generally environmental ones (e.g., Hickman and Porter 1993) – related to modern technology, but never an interpretation of globalized technology as *inherently* geared (by means of market prices) to the displacement of work and environmental loads. Within environmental and sustainability science, unfortunately, widespread doubts about ecomodernist "techno-optimism" are completely free from any social-science insights that might have been conducive to a grasp of the logic of global, sociotechnical displacement strategies (e.g., Huesemann and Huesemann 2011).

2003; Latour 2005), or the anthropology of technology and material culture (Miller 1987; Lemonnier 1993; Spyer 1998; Miller 2005). If anything, these fields dedicated to the academic analysis of material artifacts have moved even further away from overriding critical political and moral concerns with the role of technology in the contemporary world.[29] Are affluent modern people now simply too accustomed to living in a thoroughly technologized society to reflect on its specific conditions?[30] Upon browsing through the pages of recent deliberations on "technoscience" or "materiality"[31] it soon becomes evident that the asymmetric global resource flows that ultimately condition these myriad local engagements with artifacts are not matters of concern for mainstream social theories of technology and material culture.

How can we account for this? I can think of three contributing factors:

1. The conceptual constraints of the case study approach continue to shape our discourse on technology. Regardless of discipline, we tend to assume that technologies are aspects of local social organization and should be studied in those terms. We continue to study sociotechnical systems as instances of a universal human embeddedness in an immediate world of artifacts, but do not scrutinize the category – or global societal conditions of – modern technology as such. The global social order manifested in our increasingly globalized technologies is taken for granted.

2. The concern with how material artifacts are shaped and organized by social relations and cultural conceptions – and *vice versa* – only superficially transcends dualisms such as society versus nature.

---

[29] For example, Ihde (2010: 29–30) observes that mention of "dystopian godfathers" such as Ellul, Heidegger, Mumford, and Marx have mostly disappeared from contemporary philosophy of technology. It is noteworthy that a recent, state-of-the-art, 719-page anthology titled *Philosophy of Technology: The Technological Condition* (Scharff and Dusek 2014) has reprinted 58 texts on technology from Plato to Andrew Feenberg, but *not one* of them acknowledges that the intrinsic rationale of technological systems may be to redistribute time and space among social groups.

[30] Perhaps we need to remind ourselves that, if European or American lifestyles were universalized, it would require four additional planets. To exemplify the technological implications of global inequalities, the scale of Euro-American access to inexpensive electronic devices made in China is clearly contingent on abysmal global wage differences. Even more disconcerting, more than half of humanity must survive on less than $8 per day (Rosling et al. 2018), which means having access to very little technology. Against this background, it is difficult to be enthusiastic about esoteric discourses on the complex ways in which Euro-Americans experience the privileged world of gadgets in which they are immersed.

[31] Discourses on the phenomenology and the politics of materiality and material culture appear to be completely insulated from each other.

Beyond recognizing the recursive interaction between the materiality of technology and society, a more demanding challenge is to understand global society as in significant respects materially constituted. To address the "social construction of technology" in the conventional sense established in the 1980s is to disregard the capacity of world society to reorganize and redistribute the material conditions for engineering. Only by including the operation of the world economy in the concept "social" can we grasp how modern industrial technology as a whole is "constructed" through asymmetric global exchanges of biophysical resources. Such an interdisciplinary, *sociometabolic* perspective will no doubt seem alien to most social scientists.

3. The conclusion that modern technology is an inextricable component of an obscenely unequal and exploitative world order is an offense not only against centrally positioned academic fields such as economics and engineering but ultimately also against the interests of global capitalism. It can thus safely be predicted that subtle processes of selection and exclusion will guarantee that such a perspective will never become a part of mainstream discourse. A likely reaction to the argument presented here is that its implications would simply be bewildering. But however powerful they are in determining discursive closure, such pragmatic objections are political, not theoretical.

If not simply bewildering, what might the practical implications of this theoretical argument be for our approach to modern technology? The most obvious implication is undoubtedly that both mainstream and Marxist visions of a shift to low-carbon technologies to mitigate climate change are founded on misleading assumptions about what technologies really are. When David Ricardo asserted that capital could substitute for land, he codified the outlook of wealthier nations able to displace their ecological footprints to other parts of the global system. To say, two centuries later, that solar-power infrastructures require substantial investments of capital is a convoluted way of saying that they have great indirect requirements on embodied land and labor appropriated from elsewhere (see Chapter 7). Whether it is feasible to procure such volumes of embodied resources from other parts of the world depends on relative market prices of human time and natural space. The global economy is currently propelled to almost 90 percent by fossil energy, which means that the construction of nonfossil infrastructures like hydroelectric, nuclear, and solar is itself fossil fueled to the same extent. The hype about a transition to renewable energy is an illusion because the annual expansion of such energy use remains slower than the expansion of the use of fossil fuels. Nor is the world economy "dematerializing" (Schandl et al. 2016). We very urgently need to understand why our technological visions are not being implemented. We are facing an ontological cataclysm that we cannot for long circumvent. These are uncomfortable challenges, to be sure, but to choose not to address them should not be an option.

7

Energy Technologies as Time–Space Appropriation

To reconceptualize the relation between "power" in the sense of socio-political organization and "power" in the sense of energy technologies, we need to unravel the continuities and discontinuities between our concepts of "energy" and "agency." A delineation of the significance of "agency" is discussed in Chapter 10, where it is observed that so-called posthumanists misleadingly tend to attribute agency to any kind of force, regardless of whether it is propelled by a purpose of some kind. Paradoxically, my critique of the mainstream approach to energy technologies is that it tends to disregard the human agencies underlying their capacity to redistribute workloads and environmental burdens in world society. At first glance, these two objections may seem contradictory, but they are not. In the first case, I disavow the attribution of agency to nonliving forces, while in the latter, I propose that we identify the *human* agency involved in organizing the unequal access to nonhuman energy. Both arguments challenge conceptual fallacies – either blurring the distinction between the living and the nonliving, or treating technological systems as artifacts purified of human agency – that epitomize the phenomenon of fetishism. In both cases, society is made invisible. In the former case, decontextualized entities are represented as equivalent sources of agency, and the distinction between social and natural is dissolved. In the latter case, the category of "energy" reveals itself to be a politically neutral way of talking about social machinations. Bodily work – muscular energy – is always propelled by agency. When nonmuscular energy is harnessed to substitute for bodily work, there may be sociopolitical agencies at work that are less apparent.

The Notion of Biofuels

Since the 1970s I have spent most of my spare time managing a farm on the east coast of Sweden. After having raised sheep (at most close to 190 ewes) and beef cows to keep its around 40 hectares of fields and pastures from reverting to forest, I currently find it more economically rational to approach these surfaces as huge lawns to be polished with a machine resembling a large lawn mower mounted on my tractor. Each year, vast volumes of grass are simply left rotting on the ground. I keep reminding myself that these same 40 hectares (and the waters and woods around them) a century ago provided subsistence for eight large households – around 50 people – on the farm, in addition to producing a continuous flow of foodstuffs (mostly dairy products) to the nearest urban centers. I am aware that the situation is quite similar throughout agriculturally marginal areas of Europe and North America, while millions of people in Africa and South America are malnourished and in desperate need of agricultural land. It inevitably makes me wonder about the societal and cultural processes that in some parts of the world have reduced "land" from a resource crucial to human survival to lawns that we spend our spare time pruning like golf courses. It has also made me wonder about the proposal, some years back, of planting willow (*Salix*) on agricultural land for energy production. Through what kind of convoluted rationality could there now be serious advocates of encouraging the shrub lands that our ancestors struggled so hard to keep out of their fields? What does it tell us about historical changes in European attitudes to land, and ultimately also about our cornucopian understanding of technology? Could the combustion of willow bushes really yield enough energy to compensate for the energy expenditures in the production of machines for planting, fertilizing, harvesting, transporting, chipping, and burning these shrubs, as well as the consumption of fuels and fertilizers? What is the role of money and economics in promoting such counterintuitive schemes for resource management?

The derivation of mechanical energy from biological organisms was ubiquitous prior to the Industrial Revolution. For thousands of years, the use of wind and water power (and the use of peat by the Dutch in the early modern period) were marginal complements to the ancient and fundamental dependence on human labor, draft animals, firewood, and charcoal, forms of organic energy that all implied harnessing contemporary solar radiation. Coal had been used for heating in medieval Britain and elsewhere, but except for Dutch peat the main energy sources in preindustrial manufacturing processes were renewable. No premodern person would ever have thought of food, fodder, or charcoal as "biofuels," although they represent the same principle. While occasional visionaries like Rudolf Diesel envisaged running diesel engines on vegetable oil from the colonies already a century ago, the modern concept of biofuels was conceived, after more than two hundred years of using fossil fuels, as an alternative energy source that might replace fossil fuels to mitigate climate change and substitute for them as fossil deposits become scarce. Proponents of biofuels generally do not think of them as representing a regression to preindustrial times, but both their rationale and the problems they raise revitalize the logistics and dilemmas of human life prior to the Industrial Revolution.

The turn to fossil fuels and steam technology in late-eighteenth-century Britain should not simply be understood as a Promethean breakthrough in engineering, but as propelled by domestic land constraints and contingent on global economic processes and trade relations. To derive energy from beneath the surface of the earth was a highly successful option in a landscape characterized by shortages of agricultural land, wood, and reliable watercourses within easy reach of urban centers. The development of steam technology occurred to satisfy the demands of the cotton textile industry, which in turn developed in response to the great global demand for cotton textiles. It deserves to be repeated that this demand largely derived from the Atlantic slave trade. Not only did slave traders in West Africa find that industrially produced British cotton textiles successfully competed with manually produced Indian ones in the purchase of African slaves but also the owners of American plantations required cotton clothing for their slaves (Inikori 2002; Beckert 2014). In other words, the very slaves who provided inexpensive labor for harvesting raw materials for the British cotton industry were purchased and clothed with the products of that industry. Without slavery there would have been a much smaller market for the British textile factories. The fact that the Industrial Revolution, after all, was contingent on the toil of human bodies (Rabinbach 1990; Nikiforuk 2012) gives us reason to pause and reflect on our assumptions about the conditions and implications of technological progress.

Such were the historical origins of fossil fuel technology. To that majority of people who believe that new technologies merely hinge on the successful implementation of discoveries about the physical nature of things, the most important lesson is that they also require *money*, and that this is a mystified way of saying that they constitute profitable social strategies for *redistributing* biophysical resources (such as embodied land, labor, materials, and energy) in the world-system. As an anthropologist, I must conclude that the currently mainstream conception of "technology," in not acknowledging its dependence on asymmetric resource flows, is a specific *cultural category* generated by historical developments in Europe in the eighteenth century.

In not recognizing its political-economic dimension, most modern people tend to project unrealistic hopes and expectations onto technology and engineering science. The question is generally not *if* a certain problem can be solved by engineering science, but *when*. We thus confidently await improved versions of technologies for harnessing renewable energy, convinced that there can be no *intrinsic* obstacle to running a previously fossil-fueled modern society on photovoltaic energy or biofuels. From a purely physical perspective, there may not appear to be any such intrinsic obstacles. But our energy technologies are not just physical phenomena, they are embedded in global societal exchange relations that should be just as significant for determining what is feasible and sustainable as purely physical calculations. An analysis of the prospects of biofuels as a future alternative to fossil fuels must thus necessarily be interdisciplinary.

There were tangible biophysical reasons why the historical impact of fossil fuels as a source of mechanical energy was so revolutionary. Coal, oil, and natural gas are very concentrated energy sources, embodying millions of years of solar radiation. Fossil energy made entirely new technological achievements possible – from railways to space shuttles – that simply had not been feasible using preindustrial energy sources. Harnessing them also meant not having to use significant parts of the land surface to capture solar radiation in the production of organic energy through, for instance, horse fodder and charcoal. What we know as modern society – for a long time confined to the world-system cores of Europe and North America – has for two hundred years become accustomed to a succession of technological innovations made possible by fossil fuels and the economic growth based on their combustion. Over these two centuries, mainstream views of economic development have also been conditioned by the use of fossil fuels, as evident, for instance, in the conventional understanding of land – ever since David Ricardo – as a substitutable factor of production and of agriculture as a marginal or even primitive pursuit.

Geopolitically, the British adoption of fossil-fueled steam technology was both contingent on and instrumental in generating global shifts in the flows of biophysical resources. It granted Britain "ecological relief" (Pomeranz 2000) not only by liberating large parts of its land surface from the production of fodder, firewood, and charcoal but also by providing access, through the sale of British exports abroad, to vast amounts of land, labor, and materials on other continents. Through colonialism and world trade, British export production was a strategy of not only displacing workloads to plantations and mines in the periphery but also of *environmental* load displacement. By the end of the nineteenth century, Britain had access to the produce of a land area several times that of its own national territory. With the exception of mines (but not of the miners), the land (and labor) appropriated through colonialism represented means of harvesting the organic products of contemporary sunlight. Steam technology, in other words, was a means of commercially converting fossil energy into bioenergy and materials derived from vast land surfaces processing solar radiation.

Against this background, the idea of mitigating the pernicious consequences of fossil fuels by replacing them with biofuels seems profoundly flawed. A critique of this scheme can be conducted at two distinct analytical levels. It can either focus on the structural contradictions of the proposal, which revitalize the metabolic impasse of preindustrial Britain, or on the various deleterious effects of its implementation. At the first level, the resort to bioenergy as a general replacement for fossil energy can be shown to be fundamentally unfeasible, while at the second level, the actual practice of modern bioenergy production can be shown to have highly problematic economic, political, and ecological repercussions. Many of the latter repercussions are expressions of the structural contradictions inherent in the very idea of postfossil bioenergy.

At the level of fundamental oversights, the suggestion that we can replace the use of energy representing millions of years of sunlight with that of current solar radiation does not recognize the crucial significance of the vast time-spans required to concentrate the energy in fossil fuels. To believe that the energy embodied in harvests of contemporary organisms could substitute for the energy we now derive from the fossilized remains of the entire history of organic life on Earth is simply misguided. This is, of course, not to say that energy cannot be retrieved from crops, as humans have been doing for millennia, but that there are physical and logistic constraints that will preclude humankind from deriving more than a small fraction of its current energy use from biofuels. A tangible such constraint is that there simply is not enough ecologically productive space on Earth to replace a significant share of the current use of fossil energy with biofuels, even if we do not reckon with alternative uses of land to produce food and materials (cf. Smil 2015). A less tangible but no less serious constraint is the issue of *net energy* or EROEI (Hall and Klitgaard 2012). To calculate the potential of bioenergy we must subtract the energy spent on producing, harvesting, and processing it. It is certainly physically and technically possible to produce ethanol from maize, but the question is how much energy is expended in the process, in relation to the quantity of energy that can be derived from the ethanol produced (Pimentel, Patzek, and Cecil 2007). To the extent that net energy or EROEI is very low or even negative, such energy production is feasible only as long as there is *money* directed to maintaining it, in effect subsidizing the use of maize ethanol with other sources of energy, predominantly the fossil energy that currently accounts for about 86 percent of total global energy consumption (Prieto and Hall 2013). Given the convoluted nature of such energy production – which in terms of net energy is not energy production at all[1] – it is legitimate to ask whether one hectare of ethanol maize will yield more horsepower of mechanical energy than using that hectare to produce fodder for horses.

The combustion engines through which we have harnessed fossil energy represent a link between contemporary human social metabolism and the accumulated energy sedimented through the biosphere's metabolism over the long term. These engines appear to have established a conceptual lock-in or path dependency in engineering science, founded on the assumption that a return to organic energy must nevertheless continue to be based on the technological advances of the age of fossil fuels. But combustion engines may be as awkward a means of harnessing the energy of maize as horses are as a means of harnessing that of coal or oil. The physical relations between inorganic versus organic energy sources and the feasible means of harnessing them has not changed since preindustrial times, regardless of centuries of engineering science that has assumed that revolutionary new forms of harnessing energy are around the corner. To abandon fossil fuels may mean abandoning the combustion engine as the central source of mechanical energy. It may mean having to accept that the age of fossil-fueled industrialism will have been a brief historical discontinuity, an interlude of a few centuries between two very long periods of human social development based primarily on organic energy. There is yet no reason to believe that the quantitative metabolic logistics framing the latter of these

---

[1] Strictly speaking, of course, there can be no energy "production," as energy can neither be created nor destroyed, but I here use the phrase in the conventional sense of harnessing energy for commercial use.

two periods will differ substantially from those of the former. Textbooks in engineering science do not dissolve the fundamental difference between organic and inorganic energy.

The ultimate rationale of contemporary visions of high-tech renewable energy production, whether biofuels or photovoltaic power (both based on the idea of harnessing contemporary sunlight), may be to protect engineering science from fathoming the traumatic implications of its historical dependency on the fossil fuels that world leaders at COP21 in December 2015 purportedly decided to abandon. Harnessing contemporary sunlight was precisely what our eighteenth-century ancestors were very efficient at doing. The notion that humankind shall devise technologies for harvesting sunlight with higher EROEI than direct use of the products of photosynthesis is an example of the kinds of hubris inspired by its historically recent turn to fossil energy. The sooner we realize this, the greater is the chance that our civilization will be able to organize a voluntary transition to a sustainable organic energy regime, rather than succumb to unanticipated collapse (Tainter 1988).

As with photovoltaics, I am not denying that energy can be accessed through biofuel technologies, merely that these modes of harnessing energy cannot be viewed as possible replacements for fossil energy to any significant extent. The several problematic consequences of recent experiments with biofuel production have made this abundantly clear: these schemes have been criticized for displacing poor people from the land they depend on, for leading to higher food prices and concomitant increases in rates of malnutrition, for aggravating biodiversity loss, and even for generating greenhouse gases at rates comparable to those of fossil energy production (Scharlemann and Laurance 2008). The COP21 agreement to rely on future "negative emissions" of greenhouse gases based on bioenergy with carbon capture and storage (BECCS) may require one-third of current total arable land on the planet (Williamsson 2016). No less than the fossil energy technologies it was meant to replace, a large-scale production of bioenergy will be contingent on the displacement of problems to other populations, landscapes, and generations. This is why such an energy technology is inextricably interfused with questions of justice.

Social Science Perspectives on Energy Justice

"Energy justice," like environmental justice and climate justice, is a concept that necessarily spans the conventional divide between the natural and social sciences. In the remainder of this chapter I will attempt to theorize some problematic, justice-related aspects of the widely shared consensus that human societies should be aiming for a low-carbon energy future. Our visions of a just and sustainable future will need to be grounded in a theoretical and historical grasp of both the physical and societal conditions for justice and sustainability.

After decades of almost complete insulation from social theory, the topic of energy is currently reentering and reshaping the social sciences. Rather than offering a physical substratum as a determinant of human social relations, as was the inclination in some earlier models such as those of cultural ecology,[2] several contemporary approaches explore how physical and social processes are intertwined. Such exploration necessarily entails theorizing the human use of physical matter and energy to establish and reproduce social inequalities. It is thus inextricably connected to issues of justice.

Paul Stern, Benjamin Sovacool, and Thomas Dietz (2016) have provided an overview of some social science approaches to current dilemmas of energy and climate, but their compilation omits several of the most promising and relevant perspectives currently reshaping social theory. Like much of the recent literature on "energy justice" (Sovacool and Dworkin 2015; Jenkins et al. 2016; Heffron and McCauley 2017), it remains confined to the theoretically less ambitious level of policy and decision making. To engineers and policy makers, their article seems to suggest that the main contribution that social science might have to our understanding of the global energy and climate dilemma is to investigate the conditions of human "choices" and norms regarding the adoption (or rejection) of specific energy technologies. This is a highly skewed representation of the various considerations of energy use that are currently discussed in the social sciences.

Rather than take the category of "energy" or even "technology" for granted – as an ecologically problematic and unevenly distributed but ultimately practical component of modern life – a social science approach that probes beneath the surface of markets and politics must examine energy technologies as strategies for displacing workloads and environmental burdens to other sectors of world society. To harness energy is conventionally understood as a way of putting nature to work. The discourse on energy justice crucially needs to develop conceptual and methodological tools for determining to what extent an energy

---

[2] For a retrospective history, see Watts 2015.

technology is indeed simply a way of putting nature to work, and to what extent it is a way of putting other segments of global society to work. It must thus develop a theoretical framework that illuminates the continuities between preindustrial human labor, slavery, draft animals, windmills, and watermills, on the one hand, and the metabolism of modern energy technologies such as combustion engines, hydroelectric generators, nuclear power plants, biofuels, and photovoltaic panels, on the other. When physical work is delegated from human bodies to technological systems requiring continuous inputs of natural resources, the expenditure of energy is no less intertwined with political economy than in slavery.

It can, for instance, be argued that the turn to fossil energy in the nineteenth century was inextricably connected to British colonialism, asymmetric exchange, and environmental load displacement (Pomeranz 2000; Hornborg 2013a, 2013b), and that the category of "energy" emerged historically as a politically neutral way of culturally conceptualizing such processes. It has been convincingly demonstrated that the use of fossil energy to this day in complex ways implicates political processes at various levels of scale from the local to the global (Mitchell 2011; Huber 2013; Malm 2016), and that its consequences for climate change are profoundly intertwined with issues of ecologically asymmetric exchange (Roberts and Parks 2007). Underlying all this work in social science is an ambition to reconceptualize our taken-for-granted categories relating to energy use to understand seemingly neutral, practical, and technical issues as cultural mystifications of power relations (Strauss, Rupp, and Love 2013; Tyfield and Urry 2014). To the extent that the manufacture of photovoltaic technology is conducted using low-wage labor and rare earth minerals in peripheral parts of the world-system, a shift to sustainable solar energy in the most affluent nations (cf. Prieto and Hall 2013) may be tantamount to displacing work and environmental loads to poorer countries. From a critical social science perspective, it could be argued that such distributive injustice is not merely an inadvertent consequence of the energy transition, but – as in the nineteenth-century shift to fossil energy mentioned previously – fundamental to its underlying but mystified rationale (Figure 5). A similar rationale may be identified in the European turn to sugarcane ethanol and other biofuels from the global South. If Euro-American consumers can be persuaded to shift to photovoltaic panels and ethanol fuels, it will to no small extent hinge on the market prices of the labor and land that these energy sources embody.

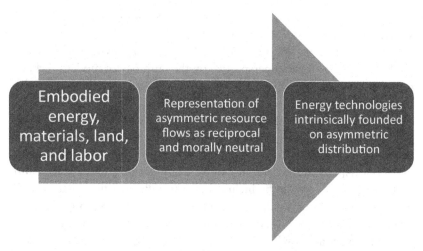

FIGURE 5 Asymmetric resource distribution as the underlying but unacknowledged rationale of modern technology

Energy Technologies as Sociotechnical Systems

A major theoretical issue raised by the concept of energy justice is how to conceptualize the relation between its natural and societal dimensions. "Energy" evokes physical forces in nature, while "justice" refers to societal principles for distribution. However, for several decades there has been a strong tendency in the human sciences to criticize the conventional Western separation of nature and society as a cultural phenomenon that has historically justified the degradation of the global environment. The movement to transcend such dualism culminated in Latour's (1993a) proposal to completely abandon the categories of nature and society. His influential claim is that this way of categorizing the world into two mutually exclusive domains has been fundamental to Western modernity since the Enlightenment. Latour argues that the distinction has been conducive to the colonial domination of peoples and landscapes and to environmental and political crises. Although his reasoning is frequently difficult to follow, a basic outline of his argument has been widely accepted in the human sciences. An implication that has been particularly influential in the field of Science and Technology Studies (STS) is that what has been perceived as scientific and technological discoveries of intrinsic and incontrovertible features of nature are contingent constructions imbued with social processes. In this sense, the natural and the social are thus inextricably interfused. Such an approach would underscore the argument that energy technologies represent social strategies no less than revelations of nature.

The discourse on energy justice brings concerns with social justice closer to a global societal perspective on the ontology of modern technology. Miller's (2014) concept of "socio-energy systems" highlights that energy technologies are social phenomena, reminding us that the conventional STS notion of "sociotechnical systems" (Bijker, Hughes, and Pinch 1987) has been unduly limited to the sociocultural construction and implications of particular technologies, without including their material, *metabolic* aspects in the definition of "social." In short, we need to realize that energy flows should be included in our understanding of society. In the nineteenth century, there was a mutual conceptual exchange between physics and economics (Mirowski 1989). While Marxist political economy intuitively acknowledges labor power as a kind of energy (Burkett 2005), for instance, early physicists conversely adopted the concept of "work" from political economy. Such long-abandoned aspirations to merge our models of nature and society within a common theoretical framework urgently need to be resumed, if we are to grasp the complex relations between energy policy, climate change, global inequalities, and financial crisis. To understand the global distributive aspects of energy technologies or sociotechnical systems we must include the physical, metabolic requisites of societies in the very definition of society. This will help us reconceptualize energy technologies as thoroughly *social* instruments of environmental load displacement. As I suggested in the previous chapter, the ultimate sociotechnical system is nothing less than the world economy.

It is well known that the quantities of energy dissipated per person vary enormously between individuals of different countries and classes, generally reflecting obscenely contrasting levels of purchasing power (WWF 2016: 74–81). The great global disparities in resource use (including energy) are generated and reproduced through the joint operation of the world market and globalized technologies. The phenomenon of modern technology inaugurated with the Industrial Revolution, generally perceived as a natural progression, was made possible by the establishment of the disembedded global market economy, by means of which the asymmetric resource flows of the British Empire could continue beyond the official end of colonialism. To this day, the wealthiest and technologically most advanced parts of the world are net importers of biophysical resources such as embodied labor, embodied land, embodied energy, and embodied materials, yet these net resource transfers remain invisible to neoclassical economics preoccupied with monetary flows and market equilibrium. Not only are the asymmetric flows of material resources culturally invisible, but concomitantly also the operation of modern technology as an apparatus for orchestrating such flows.

Energy Justice, Space, and Time

The recent discourse on energy justice tends to approach what Miller (2014) calls "socio-energy systems" as arrangements that *inadvertently* have distributive consequences that can be analyzed as injustices. Such unjust distributive consequences are presented as incidental while their identification and examination as problems of "energy justice" are portrayed as an innovative theoretical project. An alternative perspective, based on the approach to technology and asymmetric exchange offered here, is that the distributive logic of energy technologies is *internal*, rather than external, to the technologies. In this view, the distributive implications of a given energy technology, in terms of appropriation and displacement, are not incidental but central to its societal rationale. However, the individual incentives that propel the behavior patterns that generate such distributive effects are framed in the convoluted terms of market prices, which means that processes of displacement and time-space appropriation are not immediately apparent.

The discourse on energy justice tends to assume that injustices associated with energy technologies should be approached as problems of engineering rather than social science. The stance is cognate to deliberations on the "unintended consequences" of energy technologies (Andersen 2013). But what is "intended" is an ambiguous question in relation to the underlying structural rationale of social organization. Even Ulrich Beck's (1992) influential contribution to sociology focuses on the sociological *implications* of increasing technological complexity, rather than on understanding such complexity in itself as a *sociological* phenomenon of risk displacement. These approaches reproduce the conventional assumption that the physical and the technological are to be regarded as external to the social. By and large, the social sciences have relegated energy and material flows to a "natural" domain that is imagined to be external to their field of interest.

The understanding that a shift to renewable energy technologies would imply a return of land constraints – that is, that energy and space are transposable parameters – is receiving increasing attention in several disciplines (Bridge et al. 2013; Hornborg 2013b; Huber 2015; Smil 2015;

Huber and McCarthy 2017). One conclusion regarding this transposability of energy and space is that, in preindustrial societies, eco-productive space provided access to energy, while in industrial ones, fossil energy provides access to space. Prior to the use of fossil fuels, the primary source of energy was eco-productive space, which provided food and animal fodder for muscular work as well as firewood and charcoal. The turn to fossil energy reversed the relation between energy and space, as fossil fuels now propelled new transport technologies that provided access to increasingly wider spans of space. The analytical link between space and energy is time, as it requires vast spans of time for eco-productive space to yield fossil energy, and as it is by reducing the expenditure of time in movement – that is, increasing velocity – that fossil energy can provide access to more space. By increasing the speed of movement, new transport technologies propelled by fossil fuels – initially railways and steamboats and later automobiles, cargo ships, and airplanes – have made it possible for privileged sectors of the world economy to access resources and markets at increasingly greater distances and thus within increasingly wider spaces.

The theoretical integration of distributive aspects of energy, space, and time is facilitated by referring to the "time geography" developed by the geographer Torsten Hägerstrand (cf. Gren 2009). Velocity is a measure of the amount of time required to traverse a given space, and, given a certain mass and amount of friction, it can be physically expressed as the dissipation of a given quantity of energy. Applying Hägerstrand's insights on time geography, it can be argued that the saving of human time represented by the increasing velocities of fossil-fueled transport technologies is made possible through a conversion of the geological time embodied in fossil energy.

In a related argument, David Harvey (1989) has characterized modern urban life as a condition of "time-space compression," and it is obvious that the increasing velocities and obviation of spatial distances that are fundamental to this condition are contingent on an increasing dissipation of energy. Harvey's time-space compression is a result of the operation of modern technology, and it would be impossible without what I have called *time-space appropriation*. While the former concept primarily refers to human experience, the latter refers to tangible flows of material resources. As Harvey is a prominent exponent of historical materialism, he would no doubt agree that the phenomenology of time-space compression has a concrete foundation in material resource flows. The concept of time-space compression refers to how technology saves time and space for those who have access to it, while ignoring that time and space are lost for others who do not. It is noteworthy that not even Marxist discourse on capital accumulation engages the materiality of asymmetric resource flows, such as the quantifiable global transfers of embodied labor (Simas, Wood, and Hertwich 2015) or embodied land (Lenzen et al. 2012, 2013; Yu, Feng, and Hubacek 2013). The different contexts of the two arguments on time–space compression and time–space appropriation highlight the problematic disciplinary divide between the subjective and the objective, the semiotic and the material, although they are obviously two sides of the same coin.

An important component in a theoretical framework for conceptualizing energy justice as an issue representing the confluence of nature and society is the connection between the design of money and the organization of what Harvey refers to as "space–time." Drawing on Munn's (1986) study of the exchange of valuables among people inhabiting islands off the northeast coast of Papua New Guinea, Harvey (1996) observes that the specific notions of reciprocity inscribed in particular, traded artifacts generates different social constructions of space and time. Extrapolated onto modern society, this means that general purpose money establishes spatial and temporal scales of reciprocity – and thus also ranges of personal agency – that differ significantly from those generated by the exchange of so-called primitive valuables in Melanesia. This is another way of observing that modern money makes it possible for its possessors to appropriate energy and materials from much wider geographical areas – and to defer and accumulate purchasing power for greater lengths of time – than was feasible in premodern societies. The design of money artifacts is thus a societal determinant of the feasibility of particular kinds of technologies.

Another aspect of how money shapes the social perception of space and time is its capacity to represent energy as time. In modern, capitalist economies, the commodity of labor power is measured and priced in terms of expended time, although its essence is energy (Burkett 2005). The technological augmentation of labor productivity – generally reflected in higher wages – is essentially an increase in the amount of energy dissipated per hour of labor time. The accumulation of more powerful "productive forces" – technological progress – thus has cognate implications for labor productivity and velocity: both entail an increase in *exosomatic* energy dissipation per unit of time. Both are also ultimately contingent on money. A measure of a commodity's social exchange-value, rather than of biophysical inputs in its production, money has a notorious capacity to conceal – and to orchestrate – asymmetric resource flows on the world market.

Hägerstrand reminded us that both matter and movement are spatial phenomena. It is perhaps tautological to observe, as he did, that matter must always occupy space, and that movement is an activity implicating space, but an intriguing observation of his time geography is that time is defined by the existence and movement of matter, and that access to space is determined by the expenditure of time, that is, velocity. What needs to be added is that movement requires energy, and that the dissipation of energy will be proportional to velocity.[3] Genuine sustainability and energy justice would both demand constraints on the mobility of matter, including people. This is particularly evident when we consider transport technologies such as aviation and automobiles. Money and technology are social arrangements for providing some people with the capacity to dissipate more energy than others, which to a large extent entails movement in the form of travel as well as transports of commodities. They are artifacts for distributing natural resources so as to reproduce social inequalities. Although drawing on natural resources, energy technologies are thus inherently social. Given that technologies for the dissipation of energy as well as inequalities in such dissipation are contingent on relative purchasing power, efforts to achieve sustainability and energy justice – and a reorganization of global space–time – must entail a transformation of the very logic of money. Given more immediate concerns with the design of sustainable energy technologies, a primary consideration of energy justice is the extent to which the feasibility of a particular low-carbon technology, while contingent on purely monetary calculations, would be physically dependent on the asymmetric social transfer of embodied human time and/or natural space. Such "indirect" requirements of capital accumulation are denoted by the horizontal arrows in Figure 6, whereas the vertical arrows represent what is conventionally and misleadingly thought to be the physically sufficient, purely "natural"

---

[3] Access to high-energy, high-velocity technologies is very unevenly distributed in world society. It is another matter that transport technologies such as the automobile may reach thresholds of rationality from which point they are clearly counterproductive. This is illustrated by Ivan Illich's influential argument on the "real speed" of the automobile, calculated by dividing the annual distance covered by the time spent on maintaining it, paying for it, and so on. Such observations on diminishing returns on inputs, which recur in Tainter (1988), Daly (1996), and Hall and Klitgaard (2011), should be connected to the question of how inputs and outputs are socially distributed; for instance, *whose* time is spent – and at which price – and whose is saved?

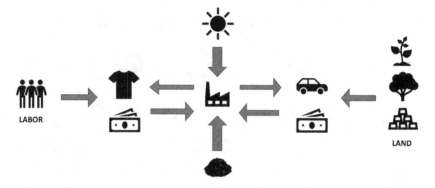

FIGURE 6 Industrial technology as capital accumulation: how energy sources obscure social exchange relations

requisites of a given technology (Figure 6). What we conceptualize as "technology" is the intersection of social and natural phenomena, namely specific, socially organized ratios of exchange and biophysical features of the material world. The technological harnessing of energy is *contingent* on the social accumulation of capital through asymmetric exchange of biophysical resources.

This can be illustrated by the unacknowledged, indirect land requirements represented by investments of labor and capital in photovoltaic technologies for harnessing solar power, which are vastly greater than the land areas occupied by the technological infrastructure (Figure 7). The indirect land requirements of labor investments can be calculated by multiplying the number of requisite person-year equivalents by the workers' average ecological footprint. The indirect land requirements represented by investments of capital are more difficult to estimate, but, as illustrated by the role of colonial plantations in generating capital investments in steam technology in nineteenth-century Britain, capital tends to signify profits accumulated from economic processes ultimately demanding expanses of geographical space. An incontrovertible fact, furthermore, is that money generated by an economy propelled by fossil energy must implicate a significant carbon footprint.

Fernand Braudel (1992: 65) observed that "there have always been a number of privileged persons (of various kinds) who have managed to heap on to other shoulders the wearisome tasks necessary for the life of all." When the toil of slaves was shifted on to watermills in the fifth century AD (Debeir, Deléage, and Hémery 1991: 39), it may have given rise to the illusion that technological progress did not implicate political

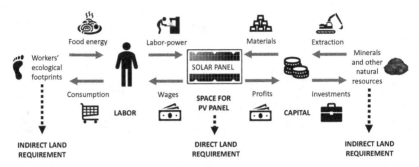

FIGURE 7 Indirect land requirements of photovoltaic technology in addition to space occupied by infrastructure

economy, as did slavery. However, to control what Marx called the productive forces or means of production is a fetishized way of appropriating *embodied* labor power as well as other biophysical resources. No less than in slavery, other modes of energy use implicate political economy as part and parcel of what historians have perceived as technological progress. To understand the transition from slavery to watermills as a phenomenon of political economy, relevant questions include: By means of which social, political, and economic arrangements were the mills built? Who owned them and reaped profits from them? How was access to adequate sites distributed? Similarly, we should rethink the Industrial Revolution – and the global role of fossil fuels ever since – in terms of the asymmetric political and economic relations that gave some parts of the world access to the labor and land of other parts. Finally, as we envisage a transition to various forms of renewable, low-carbon energy systems, it is incumbent on us to ask the same perennial question: To what extent are these technologies contingent on global differences in the price of labor and land, which generate asymmetric social transfers of embodied human time and natural space? For instance, is the underlying rationale of importing sugarcane ethanol from Brazil or photovoltaic panels from China the appropriation of embodied land, labor, and materials? Only by approaching technological systems in this way can we attribute due significance to how they interfuse aspects of nature and aspects of society.

In extending our perspective on energy technologies along these lines – from a focus on their local, technical performance to a grasp of their global conditions – we are prompted to reconceptualize even seemingly incontrovertible measures of physical efficiency such as Vaclav Smil's (2015) concept of *power density*. Power density refers to the land area required per watt of energy harnessed and is expressed as W/m^2.

As Watt (W) is a measure of energy per unit of time, W/m² considers energy in relation to both time and space. The concept usefully illuminates the transposability of space and energy that I have addressed in this chapter, but it is confined to considering merely the spatial extent of the energy source and necessary technological infrastructure. It thus does not reckon with the land areas required elsewhere to make investments in that infrastructure possible. For example, Smil's (ibid., 195) observations on the uniquely "punctiform" power density of fossil fuels,[4] although valid in terms of their advantages in facilitating environmental load displacements through trade, disregard the land requirements of the capital and labor needed to invest in technologies (such as steam power) for harnessing fossil energy to begin with. In a global, interdisciplinary sense, both cotton plantations and the ecological footprints of British and Afro-American labor should be considered part of the total land area required for the harnessing of coal in nineteenth-century Britain. From this perspective, fossil fuels are not as punctiform an energy source as Smil asserts, and, unless the technologies for harnessing them are taken for granted, their power density is very much lower than in his technical calculation. The argument is formally similar to the concept of EROEI in suggesting a ratio of Energy Return on Space (EROSI).[5] Energy technologies are not just local instruments for harnessing nature, but globally accumulated embodiments of asymmetrically traded resources including labor, energy, materials, and land. The reason why we instinctively reject the idea that the indirect land requirements represented by capital accumulation should be included in the assessment of a technology's power density is that we habitually excise the physical technology "itself" – ostensibly a politically neutral harnessing of nature – from its global social context.[6] We must nevertheless keep in mind that the technological potential of a world-system core draws on resources from the entire system, and that the physical operation of energy technologies

---

[4] The advantages of the punctiform character of fossil energy sources have been emphasized by historians such as Wrigley (1988) and Sieferle (2001), and recently reiterated by geographers Huber and McCarthy (2017).

[5] Even if precise quantification in support of this argument is currently not available, the fact that investments in technological infrastructure represent profits from – and thus claims on – land areas elsewhere is incontrovertible.

[6] It is noteworthy that historical materialists such as Huber and McCarthy (2017), who refer to Smil's concept of power density to argue that a transition to renewable energy technologies would entail greater demands on land, do not consider the indirect land requirements implied by the accumulation of capital for investment in such technologies.

should not be sequestered from the global political economy of resource flows that make them possible.

Such a rethinking of the ontology of technology takes seriously the conviction frequently voiced in the human sciences that the conventional way of delineating "nature" versus "society" needs to be revised. Rather than remain the musings of some token philosophers representing the imaginative distractions of a field of "environmental humanities," forever restrained from contaminating the technical and economic practicalities of business as usual, insights on the incontrovertible interfusion of society and nature must be allowed to illuminate the impasse of a long anticipated but still elusive transition to renewable energy technologies. To explain the failure of such a transition in terms of a lack of capital is ultimately a convoluted way of saying that the material constitution of the envisaged technologies requires more space than is socially accessible.

Can We Calculate the Land Equivalent of Money?

What does money represent? Critical or heterodox economists assume that it must somehow represent something real. Marxists argue that it ideally represents a commodity's "real value" in terms of embodied labor time. Ecological economists propose that it should represent available physical resources such as energy or the value of ecosystem services. But both these kinds of approaches seem useless in an economy where the total sum of money in circulation over time has no consistent relation to the sum of embodied labor or natural resources. The abandonment of financial constraints such as the Bretton Woods gold standard or the principle of full reserve banking – according to which banks would not be able to create credit money independently of the volumes of financial assets they have – has made it increasingly obvious that money is a floating signifier without any conceivable signified. This situation means that there appears to be no limits to economic growth, but that there may be growing discrepancies between the throughput of money, on the one hand, and material parameters such as the input of labor time or the stocks of natural resources, on the other. The phenomenon of financialization is thus conducive to increasing arbitrariness in the global market pricing of labor time and natural resources, which encourages growing economic inequalities and environmental degradation.

To assess the global rationality of a certain technology, we would need to establish a land-and-labor coefficient of capital. I write "land-and-labor" to indicate that land and labor are physically interfused. The fact that each laborer has an ecological footprint means that a measure of labor time such as a "person-year equivalent" can be translated into land. Even if we subtract that part of the worker's consumption that is irrelevant to the reproduction of his or her labor power, the food energy required to reproduce labor can be translated into a two-dimensional land surface. A greater challenge is to translate capital into land requirements. We may recognize in general terms, for instance, that the capital required to accumulate steam technology in the textile factories in nineteenth-century Britain relied on vast expanses of plantations overseas, but the incalculability of this transformation of land-and-labor into industrial capital has prevented us from understanding technological progress generally as a geographical displacement process. Because money cannot be translated into land-and-labor, the rationality of technology cannot be assessed in terms of space and time. Conventional cost–benefit and input–output analyses are not likely to tell us if, for example, a given kitchen appliance finally saves global space-time or merely redistributes it. We may intuitively understand that a given technology represents the saving of human time and natural space for some social category at the expense of time and space lost to another, but the elusive character of money obstructs any aspiration to establish such displacements in any quantitatively precise way.

When an individual or a nation invests in a technological artifact, there is currently no way of knowing the ratio between space-time won and space-time lost. The money price of the artifact tells us nothing about this. We cannot estimate the relation between the total amount of labor time and natural resources invested in the manufacture and transport of the artifact, on the one hand, and the time and space saved by the consumer, on the other. An assessment of the rationality of the transaction is exclusively based on the relative purchasing power of producers and consumers on the market. A tentative and rough approximation of technological time-space appropriation can nevertheless indicate the direction for which this research should aim. The currently conceivable measures available to us are very crude. As mentioned, the land equivalents of labor can be derived from estimates of ecological footprints, but to establish the land equivalent of capital poses enormous problems. It is one thing to acknowledge, as a general and incontrovertible condition, that the accumulation of capital is based on the exploitation of natural space, but something else to establish a quantitative method for tracing such calculations. A possible, alternative option may be to base it on the fact that (1) around 85 percent of all commercial energy use in the world derives from fossil fuels and that (2) the use of fossil fuels can be attributed an average ecological footprint per unit of dissipated energy. One way of calculating the footprint of fossil energy use is to estimate the land area required for sequestration of its carbon dioxide emissions. A tentative carbon footprint per US dollar in a given year could be calculated by multiplying the global use of fossil energy (in MW) by the average carbon footprint from the combustion of fossil fuels (in hectares/ MW) and then dividing global GDP by that figure and finally multiplying by 85 percent. The resulting figure would roughly correspond to a land equivalent of money, enabling us to assess the extent to which capital investments in purportedly land-saving technologies are tantamount to displacing land requirements elsewhere.

What Does "Energy" Mean?

What is denoted by the concept of "energy"? A common answer would be that it is a physical force present in nature and a measure of the ability to conduct work. In line with this view, energy technologies such as steam engines, combustion engines, nuclear power plants, and photovoltaic panels are conceptualized as products of human inventiveness and contingent on human knowledge about nature. The prevalent understanding of such technologies is that they are means of harnessing natural forces made possible by the state of human knowledge.

But this mainstream view of energy technologies is problematic. To begin with, the definition of energy as a measure of the ability to conduct work is flawed, because even "unavailable" energy such as heat dissipated into space is unquestionably energy, although it cannot do work. The distinction between "available" (or "free") and "unavailable" energy, moreover, is not simply based on physical features, but dependent on the state of technology. Fossil, hydroelectric, nuclear, and photovoltaic energy were not "available" for human harnessing until technologies for doing so had been built. This means that the concept of "available energy" is a *relational* term that denotes the technologically mediated relation between humans and their environment in a given social context.

It is rarely acknowledged, however, that the technologies by means of which given energy sources are made available for human harnessing are contingent not only on the state of human knowledge but also on the socially organized, asymmetric transfer of biophysical resources between different categories of people. The concept of "technology," in other words, is also relational. The expansion of steam technology during the Industrial Revolution, for instance, was contingent not only on some

British inventions but also on the British net appropriation of embodied labor-power, land, energy, and materials from the peripheries of the eighteenth- and nineteenth-century world-system. As all these biophysical resources represent some measure of embodied, horizontal space, energy technologies may be assessed in terms not only of "energy returns on energy invested" but ultimately also in terms of "energy returns on space." The vertical extraction of fossil energy did not make eco-productive land superfluous to the economy but reinforced and extended British access to the peripheral land areas that had made the accumulation of steam technology possible to begin with. The category of "technology," in other words, should not be understood simply as referring to the revelation of forces present in nature, but as a social strategy for utilizing natural phenomena to generate and reinforce social inequalities.

Energy technologies prompt us to reexamine our conventional distinction between the natural and the social. They are socionatural phenomena but we tend to be convinced that they refer to and presuppose a politically neutral domain of reality waiting to be more fully revealed. There is unquestionably an objective framework that defines the ultimate limits for what humans could potentially accomplish, but the role of the increasingly globalized economy in making various technologies feasible demonstrates that this framework is not exclusively "natural." The global economy interfuses the physical potential of nature with the redistributive potential of world society, enabling privileged categories of people to travel in airplanes or space shuttles fueled with miraculously available, increasingly "unconventional" fossil energy.

Nor is the state of knowledge about harnessing natural energy a progression in the sense that new inventions are added onto previous ones, as earlier technologies (such as those for using draft animals in agriculture and transports, or wind power for windmills and sailing ships) tend to be abandoned and unavailable for most of the population. If the advanced technologies currently contingent on a globalized economy should fail, it would be no easy task to revert to the technologies employed by earlier generations. Modern people thus tend to be enmeshed in systems of global resource flows that are so essential to their survival that disturbances in these flows would have very serious consequences. This applies most alarmingly to their dependence on modern infrastructures for providing them with food energy.

The concept of "energy" is a cultural category conceived at the interface between physics and economics (Illich 1983; Mirowski 1989). As both human labor and the objects of consumption have become commodities on the market, they can be quantified in terms of money as well as in terms of (embodied) energy. The output of human energy and the input of consumption goods can be seen as a transformation of work, through money, into consumption, that is, labor energy – money – the energy embodied in commodities ($E - M - H\ E^1$). The abstraction and disembedding of energy into such transformations of market commodities is the point of departure for expanding cycles of consumption. The historical replacement of the ideal of low-energy efficiency (resource frugality) with the ideal of high-energy efficiency (conceived in terms of monetary cost–benefit analysis) entailed a shift to a modern rationality that generates increasing degradation of the environment and appropriation of other people's embodied labor.

The history of ideas on how technologies for harnessing inorganic energy have implied transformations of human relations to space or land can be divided into three approaches. The first focuses on the observation that the turn to fossil energy represents an emancipation from land (Wrigley 1988; Sieferle 2001). It was conducive to Ricardo's conclusion that other factors of production – primarily capital – may substitute for land, and it has been foundational to the emergence of neoclassical economics. The second acknowledges that a postfossil, renewable energy regime would be tantamount to a return of land constraints (Smil 2015). This conclusion is consonant with the conviction that land is nonsubstitutable, as is held in the field of ecological economics. Finally, a third position would argue, as I have done in this chapter, that *all* modern energy technologies represent accumulations of capital that ultimately require land and labor elsewhere. "Capital" represents *other* people's land and labor. In this view, energy technologies are instruments for redistributing space and time in world society. Such a view, which recognizes how money, energy, and space are transposable through technology, might be called "topological" economics.[7]

---

[7] A key theme of topology – popularly referred to as "rubber sheet geometry" – is that seemingly disparate phenomena can be understood as transformations of each other that are made possible by certain nonarbitrary principles or algorithms. For most anthropologists, such structuralist perspectives are applicable to semiotic phenomena, but not to the material world to which they refer. In this chapter, by contrast, I have argued that technologies are ultimately links in transformative sequences by which space, by means of capital, is converted into energy. In theory, the conditions for such transformations should be possible to specify in quantitative terms. For instance, how many hectares of plantation land and hours of labor, at a specific cost in relation to the market price of cotton textiles at a given point in time, will yield the sum of capital required to invest in a certain scale of steam technology?

8

Capitalism, Energy, and the Logic of Money

To most people, money is like water is to fish – and yet it is a most peculiar invention. In this and the following chapter, I will elaborate my argument that its intrinsic logic is nothing less than *capitalism*. Marx seems to have been profoundly aware of this,[1] which raises the question of why neither he nor most of his followers made the very artifact of money the primary target for reform.[2] Although Marx and a minority of Marxists envisaged postcapitalist society as necessarily moneyless, such visions have universally been dismissed by practical politicians as utopian (Nelson 2001a). In contrast to so-called utopian socialists like Robert Owen, political regimes inspired by Marxist theory have generally not contemplated radically transforming the role of money.[3] Yet money in its present form inexorably generates not only increasing global inequalities but also an appalling degradation of the biosphere. The basic conundrum of money is that it is simultaneously an idea, or *sign* (a unit of account), and a potent material force. Money is represented as a reflection of some underlying and more material level of reality to which it refers, yet it

---

[1] As with so many other aspects of Marx's creatively "undisciplined" thinking, his understanding of money has become the focus of exegesis and voluminous academic deliberation (e.g., Sieber [1871] 2001; de Brunhoff [1973] 2015; Nelson 1999, 2001b; Moseley 2005, 2016).

[2] As we shall see, precisely the same question can be raised regarding Marxian approaches to technologies.

[3] Serious visions of a moneyless economy were nevertheless integral to Bolshevik discussions in Russia in the years following the revolution in 1917, as well as to debates in Cuba in 1963–1965 (Nelson 2001a: 505–510). A prominent proponent of the abolition of money was Che Guevara, whose decision to leave Cuba was prompted by his defeat in the debate over this issue (ibid.).

organizes that material reality. This duality explains Marx's contradict-ory approach to money, as evident in his critique of utopian socialists like the Owenites, whose ambition to abolish money he dismissed as "useless and idealistic" (Nelson 2001b: 46). Rather than seeing money as the source and essence of capitalist property relations, Marx argued that money could only be dethroned by first transforming the social relations of production.

Instead of immersing ourselves in libraries of abstract theoretical argu-ments on all the possible complexities of the logic of capitalism – whether viewed from a bourgeois or Marxian perspective[4] – we need to take a detached, second look at the seemingly self-evident, cultural idea of general purpose money, namely the idea that everything is interchange-able on the same market. This idea, embodied in money tokens, makes it possible to purchase human time as well as entire ecosystems as market commodities. This prompts everyone to accumulate as many abstract claims on other people and resources as they can, which implies wanting to pay as little money as they can for what they buy. The market will thus favor production processes with as low wages and as scant environmental considerations as possible. This logic is the common denominator behind slavery and climate change. If we don't question the idea of money, both slavery and environmental degradation will be understood as conse-quences of "underpayment" – whether underpayment of labor or under-payment of so-called ecosystem services – but the notion of underpayment only makes sense as long as we remain confined within the conceptual universe of general purpose money, which assumes that everything has a correct price. It implicitly accepts the ideological illusion that in principle all things can be exchanged at rates that can be objectively established as equal and fair. The illusion of abstract equivalence among incomparable qualities – the very foundation of capitalist social organization – is as misleading in terms of its implications for social justice as it is in terms of ecological sustainability (Nelson 2001a). As we shall see in this and the following chapter, this illusion is encapsulated in the universally employed but insidious concept of *value*.

The logic of capitalism has been intricately elaborated by both bour-geois and Marxist economists, all similarly eager to discover the covert regularities and trajectories of societies organized in terms of general

---

[4] Note that this would include most of the literature on economics since the mid-nineteenth century. Whether bourgeois or Marxist, this literature shares two fundamental features: its conceptual commitment to the idea of money and its esotericism.

purpose money. The complex implications that can be derived from the rather simple idea of such money have preoccupied vast numbers of theorists and mathematicians united by the assumption that this social invention is as immutable as natural law. The potential repercussions of various economic policies and regulations have been traced and debated by countless people committed to the conviction that there is no alternative to general purpose money. As traditional world-system cores in the 1970s faced economic decline and fiscal deficits, the so-called neoliberal program emerged as a reflection of the imperatives for efficient capital accumulation inherent in the logic of modern money.

Regardless of how repulsive we find the prescriptions of neoliberal economists, they appear to have spelled out the polarizing logic inscribed in the idea of general purpose money. In this convoluted sense we may have to concede that the policies advocated by Margaret Thatcher and Ronald Reagan – and no less the trajectories of post-Maoist China and post-Soviet Russia – reflect some fundamental requirements on the management of such money. Given how the conceptual constraints of general purpose money generate a specific definition of market "efficiency," perhaps mainstream economists cannot be blamed for endorsing economic policies that increase the abysmal global inequalities and threaten to make the planet uninhabitable for human beings. The absurdity of allowing a specific human artifact to drive our species to extinction must clearly remain beyond the pale for people whose daily activities and sense of meaning hinge on remaining within the conceptual constraints provided by this artifact. Like the ancient Easter Islanders chipping away at their stone statues, we are prevented by our cultural priorities from dealing with the fact that we are destroying our ecological context.

Although to some theorists it may sound impermissibly reductionist, most of the ills of the system that we refer to as capitalism can be derived from a simple logic inherent in the artifact (or meme) of money. It is the idea of money that makes it possible to purchase and own labor and land, and the commodification of labor and land are, of course, fundamental to capitalism (Polanyi 1944). To refer to what is commonly vilified as "capitalist property relations" without tracing them to money is to omit the core issue. When Marx captured the central element of capitalism in the formula $M - C - M^1$, his point of departure was the existence of money. And because the accumulation of money is physically boundless and infinitely attractive, humans over the centuries have devised numerous strategies for aggrandizing their monetary wealth. Alongside mercantile and industrial investments, we have recently been reminded of the potential for financial speculation, tax evasion, bribery, and a wide spectrum of corrupt and clearly criminal ways of making more money. The "human weakness" identified by the Financial Crisis Inquiry Commission (2011, xxiii) as underlying the Wall Street meltdown in 2008 is no doubt pervasive among humans, regardless of political regime. Much everyday media attention is devoted to revealing and morally denouncing the avarice of politicians, corporate executives, and other public figures. But should we really be shocked and indignant because some humans do what many of us would do if we had the chance? Greed is built into the very idea of money. As Marx ([1857] 1973) observed in *Grundrisse*, "greed as such [is] impossible without money."

To focus on the exploitative and destructive logic of the idea of money might finally help us envisage the end of capitalism.[5] To choose to scrutinize the implications of a particular human artifact rather than immerse ourselves in theoretical deliberations on the structural ramifications of a specific "mode of production," may at first sight be viewed as a concession to the radical empiricism of object-oriented ontologies such as Actor-Network Theory (Latour 2005). The great differences in outlook between Marxism and Actor-Network Theory help explain why the attribution to money of all the evils of capitalism was not compatible with Marx's project.[6] Conversely, the ideological differences between them may explain why, remarkably, leading Actor-Network theorists such as Bruno Latour have chosen not to scrutinize the most potent human artifact of all. I am not suggesting that we think of money as having agency, which would be to fetishize it, but that we acknowledge the extent to which it encourages specific inclinations in the aggregate manifestations of human agency.

If acknowledging the significance of specific artifacts is tantamount to a concession to Actor-Network Theory, it is the only concession that a Marxist political ecology would need to grant to Actor-Network

[5] Envisaging an end to capitalism is no doubt preferable to imagining the end of the world, even if it is more difficult.
[6] As Nelson (2001b: 60–61) observes, "Marx is mainly engaged in criticizing, correcting and perfecting abstract theories, despite his claims to being a materialist, which one might expect to imply a more empirical bent."

Theory. As I argue in Chapter 10, artifacts may have fundamentally transformative social *consequences*, but never agency. A truly critical social theory will need to retain a realist ontology and robust analytical distinctions between the symbolic phenomena deriving from human society, on the one hand, and the nonsymbolic aspects deriving from prehuman substrates of the universe, on the other. Posthumanist attempts to dismantle all distinctions between the social and the natural, humans and nonhumans, and even subjects and objects, will not help us. To dissolve such distinctions is not to challenge power hierarchies – on the contrary, it is to abandon our capacity to do so. Fortunately, most eco-Marxists stay away from the so-called post-Cartesian deliberations of the posthumanists. A coherent political-ecological understanding of the Anthropocene can only be built on the analytical distinction between the societal logic of capitalism (i.e., money), on the one hand, and the presymbolic and nonhuman aspects of the biosphere revealed by sciences such as thermodynamics, geochemistry, and ecology, on the other. Only by recognizing this distinction can we identify the potential for human choice and the extent of political responsibility.

To attribute a pervasive social causality to a specific human artifact does not mean subscribing to Actor-Network Theory or more generally to a posthumanist outlook. The recent discovery that human society is transforming planetary carbon cycles should least of all lead to a dismantling of social science. The notion of the Anthropocene risks making a social process look like the inevitable product of human biology. It is obvious that no other species could have invented capitalism, but as Andreas Malm and I have argued (Malm and Hornborg 2014), we must keep in mind that the Anthropocene is the creation of a minority of the human species in its struggle to dominate and exploit the global majority. It was to make this point that Malm in 2009 coined the currently influential concept of the Capitalocene.

To transcend the impasse of the Anthropocene, we must avoid ideological dogmatism and disciplinary myopia, that is, the automatic rejection of unfamiliar and undigested thoughts. Like other movements and faiths, Marxism has suffered from processes of internal contradiction, exclusion, and fragmentation. There are thus obvious analogies between the historical development of Marxism and that of Christianity. Although Marxism and Christianity are often viewed as opposed to each other, the analogy between them is not far-fetched. Both aspire to provide a holistic worldview embodying not only analysis but also moral and political imperatives and visions of the future. Both are fundamentally committed to justice and to the compulsion to challenge illegitimate power and inequalities. Both are based on the revelations and subversive proposals of venerated founding figures. And both may belong to the same millennial tradition of ideas challenging the deification of money. The Christian socialist R. H. Tawney (1972) traced a genealogy from St. Paul and Thomas Aquinas to Karl Marx. Indeed, Pope Francis's (2015) encyclical *Laudato si'* expresses many of the convictions and prescriptions of political ecology.

Considering a Marxist Approach to Ecologically Unequal Exchange

John Bellamy Foster is widely recognized as a leading proponent of what is frequently referred to as ecological Marxism, or eco-Marxism for short. His special challenge has been to prove that Karl Marx was not at all the Promethean, technological modernist that he has generally been perceived as (e.g., Benton 1989), but rather highly concerned about environmental issues and, in fact, "one of the founding figures of ecological economics" (Foster and Burkett 2008: 3, 27). In an article co-authored with Hannah Holleman (Foster and Holleman 2014), Foster similarly sets out to rewrite the history of theories of ecologically unequal exchange, presenting them not as complementary but as integral to Marx's original thought on capitalist exploitation. Foster's ambition to ecologize Marx is of course laudable, but it is important to clarify some of the analytical problems with applying a classical Marxian approach to the environmental inequalities generated by the world economy. These problems are of crucial significance to the wider argument in this book because contemporary concerns with global environmental justice are frequently couched in the conceptual framework of Marxism, whereas that framework tends to suffer from what I refer to as *machine fetishism*. To unravel these analytical problems, I shall devote the remainder of this chapter and the next to eco-Marxist theory.

The gist of Foster and Holleman's argument is that the systems ecologist Howard T. Odum's (1988, 1996; Odum and Arding 1991) approach to ecological unequal exchange was compatible with, if not inspired by, the Marxian approach to "economic" unequal exchange. Foster and Holleman reiterate the Marxian conviction that unequal exchange is an asymmetric transfer of *value*, traditionally conceived in terms of embodied abstract labor but, thanks to Odum, now expanded to include embodied energy. In their view, embodied labor and embodied energy constitute material use-values that are underpaid by their exchange-values on the capitalist world market.[7] The resulting, asymmetric flows of these use-values to centers of capital accumulation are the essence of capitalist exploitation and profits. The materialist foundation of Marxian economics is particularly clear in Foster's earlier conclusions that labor power is "an energy subsidy for the capitalist" (Foster and Burkett 2008: 6) and that Marx had presented an "energy income and expenditure approach to surplus value" (Burkett and Foster 2006: 126). As, according to Marx, "labor-power itself is energy" (ibid., 120), it should not be surprising that there are affinities between theories of unequal exchange focusing, respectively, on labor and energy.

As environmental justice activists all over the world have discovered, there is a structural tendency toward ecologically unequal exchange of resources on the world market. Asymmetric transfers of embodied energy, materials, and land are systematically concealed by the apparent reciprocity of market prices. This has been abundantly demonstrated, for instance, by numerous analyses of energy flows, material flows, and ecological footprints.[8] A crucial question, however, is if the identification and discussion of ecologically unequal exchange is best conducted in terms of asymmetric flows of values or in terms of asymmetric flows of biophysical resources.[9] To refer to asymmetric flows of energy, materials, or embodied land as an unequal exchange of "use-values" is to confuse physics and economics. The concept of use-value could be expected to refer to what people find useful, or what conventional economics calls *utility*. It is thus defined by the cultural semiotics of consumption and cannot be measured in biophysical metrics such as joules, tons, or hectares. Such metrics, however, are our only means of demonstrating the occurrence of ecologically unequal exchange. Foster and Holleman's

---

[7] Emmanuel (1972) applied the Marxian notion of labor value to international trade, arguing that the values embodied in exports from low-wage countries are underpaid in relation to those from high-wage countries, a pattern that generates asymmetric transfers of surplus value on the world market. Foster and Holleman (2014: 29) similarly define *ecologically unequal exchange* as "the disproportionate and *undercompensated* transfer of matter and energy from the periphery to the core" (emphasis added).

[8] Foster and Holleman's complete silence on material flow analysis and so-called Physical Trade Balances (e.g., Pérez Rincón 2006) is a major omission.

[9] The same question, of course, pertains to "economic" unequal exchange, i.e., asymmetric transfers of embodied labor: should these unequal transfers of what is conceptualized as labor value be measured in dollars or expended labor time or energy? The fact that Emmanuel (1972) and Amin (1976) calculate such flows in dollars (cf. Lonergan 1988: 135) is difficult to reconcile with Foster's and Burkett's assertions that, in Marx's view, surplus labor is a transfer of energy to the capitalist (Burkett and Foster 2006: 120, 126; Foster and Burkett 2008: 26).

notion of underpaid use-values thus fails to help us establish the material asymmetry of world trade. There is no way of deducing any kind of economic value from invested joules, tons, hectares, or hours of human labor. This conundrum is more systematically addressed in the next chapter.

The most fundamental reason why unequal exchange should be identified in biophysical terms, rather than in terms of contested notions of undercompensated value, is that the latter terminology can be seriously misleading if applied to deliberations on environmental justice. To phrase unequal exchange in terms of discrepancies between values and prices is to imply that it can be understood in terms of underpayment. But, as Georgescu-Roegen showed, money cannot compensate for entropy, whatever the mode of production. A clear analytical distinction between the physical and the economic makes it easier to assess the prospects and objectives of struggles for environmental justice.

Energy, Money, and Technology in Marxist Thought

In approaching the relation of Marxist theory to energy, it is useful to consider the approach known as *political ecology*. The school of political ecology initially emerged in the 1970s and 1980s among anthropologists and geographers who were dissatisfied with the functionalist assumptions of the school of cultural ecology. In the 1960s cultural ecology had focused on case studies of traditional subsistence practices in local communities as adaptations to the nonhuman environment. Pioneers of political ecology such as Eric Wolf,[10] Harold Brookfield, and Piers Blaikie rejected functionalism and instead emphasized the role of power inequalities and global political economy in shaping human-environmental relations. They demonstrated how the local resource flows studied by cultural ecology were connected to the asymmetric metabolic flows of the world-system.

It is a paradox that both political ecology and the cultural ecology to which it was opposed frequently refer to Marx. The founding figures of cultural ecology, such as Julian Steward and Leslie White, were heavily influenced by Marxist materialism. Although rarely viewed from a global perspective, energy flows were a central concern of cultural ecology. This concern with energy was Marxist in the crude sense of

[10] See Wolf 1972.

aspiring to explain cultural superstructures as reflections of material infrastructures, but for cultural ecology this approach had become a functionalist concern with adaptation to the natural environment. Political ecology, while sharing the focus on the environment and natural resources, looks for explanations of human-environmental relations not in the local ecosystem, but in the global economic system that ultimately conditions local life. The energy flows that concern political ecology are not so much those of the local ecosystem as of the capitalist world-system. The common concern with energy present in both schools, however, leads us to ask precisely what the relation is between Marxist theory and thermodynamics.

Although Marx and Engels were not persuaded by Sergei Podolinsky's ([1883] 2008) proposal in 1880 that the theory of surplus value could be phrased in terms of energy (Martinez-Alier 1987), and although the eco-Marxists John Bellamy Foster and Paul Burkett have consistently defended Engels's dismissal of Podolinsky as an "energy reductionist" (Foster and Burkett 2004; Burkett and Foster 2006), there are reasons to believe that what Marx had in mind in his deliberations on the exploitation of labor power was something quite close to the concept of energy. What Podolinsky recognized was that Marx's central conviction – that economic value derives from the material agency of labor power – suggests an intuitive attempt to integrate economics and thermodynamics.

Energy is indeed a common denominator of the various kinds of unequal exchange and asymmetric resource transfers that have been requisite to different modes of capital accumulation throughout history. Conceived as material infrastructure, the accumulation, maintenance, and reproduction of capital is as contingent on a net appropriation of energy as is the survival of any biological system. Without a continuous import of energy, a machine is as inanimate as an organism that has starved to death. These are facts of physics, but they cannot be translated into economics. Energy is expended – that is, dissipated – in any production process, but the amount of energy expended cannot be translated into exchange-value. In other words, exchange-value cannot be analytically derived from the amount of energy expended. Value is a concept deriving from the market, and the only conceivable metric for measuring it is money. Again, regardless of whether it is provided by draft animals, human labor, or fossil fuels, energy is *not* a value. It is measured in joules, not money.

Marx's insights on the social repercussions of money did not prompt him to advocate a transformation of the money artifact, perhaps because the notion that specific artifacts can generate particular forms of social organization was not a prevalent component of European worldviews in the nineteenth century, or maybe because the prospects of transforming money seemed less realistic than a revolution. In most modern Marxist thought it seems to be held that money – which, inspired by Shakespeare, Marx called the "universal whore" (Marx 1844) – could be extricated from its capitalist context and usefully deployed in the organization of a postcapitalist society. A similarly ambivalent stance characterizes Marxist relations to capitalist technology. Although Marx clearly understood the machines of his time as devices for exploitation, he believed that they too could be extricated from capitalism and made to serve the proletariat. Paradoxically, his point that technologies generate specific forms of social organization – as in his classical observation that "the steam-mill [gives you] society with the industrial capitalist" (Marx 1847) – can be viewed as a precursor to Actor-Network Theory,[11] yet he must have visualized capitalist machines as intrinsically nonsocial products of engineering, that is, as "productive forces" detachable from their social context. This is a paradox because it means that the theorist who taught us to understand how artifacts in capitalist society tend to be fetishized – that is, to represent social relations as if they were relations between things – was suffering from the illusion that I have called machine fetishism.

Most modern technology would not exist without general purpose money.[12] Modern technological systems tend to be made possible by the differences in how human time and natural space are priced in different parts of the world. They are tantamount to displacements, by means of unequal exchange rates on the market, of work and environmental burdens to groups with less purchasing power. Globalized technologies, or symbiotic constellations of money and technology, are the most recent addition to the historical succession of exploitative social arrangements beginning with slavery and serfdom. And yet modern technology is exactly what the ecomodernists hope will save us from economic and ecological collapse. The *Ecomodernist Manifesto*

---

[11] Hylton White (2013: 679–680) argues that Marx was indeed "the theorist of a sort of actor-network," whereas Latour, "by refusing to theorize social form as such,... ends up simply replicating the way that the society of the fetishism of commodities presents itself: as the only way of life we can possibly have."

[12] In a nutshell, this argument on the modern money-energy-technology complex can be phrased in the form of a syllogism: if technology is a matter of access to energy, and access to energy is a matter of money, then technology is a matter of money (Hornborg 1998a: 132).

(Asafu-Adjaye et al. 2015) is a paradigmatic illustration of the kind of technological fetishism that pervades so much of the contemporary deliberations on sustainability. It seems completely oblivious of the fact that it is precisely the globally uneven accumulation of technology that has brought us to the brink of collapse. Globalized technologies are ways of saving time and space for some at the expense of time and space lost to others. The asymmetric global flows of resources are what make modern technologies possible, and those flows are orchestrated by money. When we refer to the existence of a certain kind of technology, we tend to equate it with a corpus of know-how, a state of engineering, a set of ideas about how something can be achieved. This cognitive aspect of technology is no doubt a necessary condition for its existence, but it is not a *sufficient* condition. A requisite of modern technology as a physical phenomenon is an unequal or asymmetric societal exchange of resources such as embodied labor, energy, land, or materials. I am not referring here to technology conceived of merely as blueprints produced by inventors, but lacking material manifestation. Such ideas of engineering do not deserve to be called technology any more than a representation of a DNA molecule deserves to be called an organism. Technical blueprints and genetic codes are abstractions, and both can prove to be unviable in real life. Both technological and biological systems are unfeasible without specific structures of material exchange with their environments. Just as the existence of the organism is contingent on certain flows of energy, water, oxygen, and so on, the existence of a technological system is contingent on specific inputs of energy and materials. These latter flows are organized by the economy. Technologies are thus contingent on the ratios at which energy, materials, and other inputs and outputs are exchanged in human societies. In accordance with the Second Law of Thermodynamics, we know that the output of any technological system will represent less available energy or productive potential than the input required for its production. To be viable, in other words, a technological system must be reproduced through ecologically asymmetric resource flows. To not see this dependence of technology on unequal exchange is what I have called machine fetishism.

Contrary to mainstream conceptions of technological progress as local advances in engineering that could potentially benefit all humankind, we must reconceptualize the technological infrastructures developed in wealthier parts of the world as products of accumulation based on global relations of unequal exchange with less affluent areas. Technological progress, in this view, is not only a matter of ingenious breakthroughs in engineering but also simultaneously of devising new and profitable systems for displacing work and environmental pressures to other populations and geographical areas. This is the essential rationale of globalized technological systems. Rather than an index of generalized human progress – the pure, transcendent knowledge epitomized by the myth of Prometheus – technology since the Industrial Revolution is fundamentally an arrangement for redistributing resources in global society. Most modern technology requires not just ingenuity and specialized knowledge but also global discrepancies in market prices. It is thus as inextricably connected to societal injustices as slavery or serfdom. It owes its existence to general purpose money. Like the money that engenders them, globalized technologies are inextricably *social*. The phenomenon of money cannot be grasped without recognizing its function as a mystification of unequal exchange, and the phenomenon of modern technology cannot be grasped without recognizing its reliance on money. The phenomenon of technology, in other words, cannot be grasped without recognizing its reliance on unequal exchange.

The mainstream conception of technology tends to remain immune to deconstruction by relegating the responsibility for its various deleterious political and ecological consequences to its misuse by unscrupulous corporations and politicians. There is a widespread belief that technology "as such" would not have to implicate injustices, and that this purified, cognitivist understanding of technology can be distinguished and extricated from political contexts in which it is used in such a way as to generate social injustices and ecological degradation. This belief in a decontextualized, transcendent, and emancipatory science of engineering has been fundamental to classical Marxist visions of social progress.[13] Marx appears to have been convinced that the steam engine could serve the purposes of a postcapitalist society without presupposing slavery and soil erosion in colonial cotton plantations, let alone global warming. The inexorable progress of the productive forces, according to the classical Marxist vision, would deliver humanity from drudgery. Such Promethean understandings of technologies as innocent manifestations of Enlightenment, detachable from the societal contexts in which they have emerged, share the mainstream bourgeois illusion of machine fetishism.

For more than two hundred years fossil fuels have not only been fundamental to our technologies but have also conditioned our ways of

---

[13] It is a paradox that historical materialists in this respect appear to conceive of technologies as *ideas* rather than as vortices in asymmetric societal flows of matter and energy.

thinking about economics, even in heterodox economics, without us realizing how much our conceptions of economic growth and techno-logical progress have depended on the specific properties of fossil fuels.[14] The money-energy-technology complex frames the conceptual horizons even of its critics. To envisage the road to a postcapitalist society as a matter of shifting energy sources and collectively controlling the money while basically retaining the conventional idea of money and hoping to maintain our current technologies and consumption levels is delusory. A vision of postcapitalism that retains the idea of general purpose money is a contradiction in terms. The organizational inertia of general purpose money runs counter to the ideals embodied in socialist thought. To suppress this inertia requires totalitarian regulation that runs counter to ideals of democracy and personal freedom. Rather than build society on a tension between the inertia of its artifacts and the political attempts to regulate it, it must be a better idea to design those artifacts so that their consequences are more closely aligned with political ideals.

It is understandable that a general critique of money will seem incom-prehensible to people who cannot see that the design of money is some-thing about which we have a choice, but this just illustrates how constrained our imagination tends to be by the artifacts that currently rule our lives. I cannot believe that it would be feasible to completely abolish money and markets in human societies, as advocated by propon-ents of "non-market socialism" (Rubel and Crump 1987; Nelson and Timmerman 2011), but I believe that money can be *redesigned* so that its inherent logic would be to increase diversity and sustainability, while reducing social inequalities and vulnerability. The main objective must be to distinguish between different kinds of currencies that serve morally incommensurable purposes. To illustrate such moral incommensurability, infant mortality and environmental degradation in sub-Saharan Africa must be insulated against financial speculation on Wall Street.[15] It is incontrovertibly immoral to allow one person's survival to be geared to

---

[14] It is difficult to believe that what the world's leaders decided in Paris in December 2015 was *really* to abandon what for more than two centuries has been their main source of economic and technological growth, and what continues to account for 86 percent of global energy use.

[15] The challenge of insulating "democratically controlled local flows of money from global capital" (Mellor 2009: 36) is the widely experienced imperative underlying the many experiments with alternative economies from the Owenites to the local exchange trading system (LETS). Unfortunately, they all appear ultimately to have been unsuccessful, for reasons that I elaborate in Chapter 13.

another person's strategies to accumulate extravagant wealth, as is currently the case. With the abandonment of a single and global general purpose currency, we would dismantle the very foundation of capitalism. If market actors find it in their interest to distinguish between transactions conducted within a local sphere of exchange committed to subsistence and community, on the one hand, and a global sphere of nonsubsistence exchange and communication, on the other, much of the conditions for unequal and unsustainable exchange would rapidly atrophy. Rather than a product of coercion, this would be a consequence of voluntary human agency guided by the transformed logic of their money artifacts. Whether our priority is to avoid global financial crises, increasingly obscene inequalities, or catastrophic climate change, we shall have to fundamentally redesign the operation of money. This is the only possible road to a postcapitalist society. What mainstream economists celebrate as growth is really accumulation, and what they endorse as globalization and free trade is imperialism and unequal exchange. Our modern preoccupation with money and crucially also with technological progress can be understood as fetishism, in the sense that relations between people assume the appearance of relations between things.

9

Unequal Exchange and Economic Value

In this chapter I will elaborate the critique, introduced in the previous chapter, of some conceptual flaws in eco-Marxist theory. I argue for an extended application of Karl Marx's insight that the apparent reciprocity of free market exchange is to be understood as an ideology that obscures material processes of exploitation and accumulation. Rather than to confine this insight to the worker's sale of his or her labor power for wages and basing it on the conviction that labor power is uniquely capable of generating more value than its price, it is evident that capital accumulation also relies on asymmetric transfers of several other biophysical resources such as embodied non-human energy, land, and materials. The notions of *price* and *value* serve to obscure the material history and substance of traded commodities. This shift of perspective extends Marx's foundational critique of mainstream economics by focusing on the unacknowledged role of ecologically unequal exchange but requires a critical rethinking of the concept of "use-value." It also requires a fundamental reconceptualization of the ontology of technological progress, frequently celebrated in Marxist theory.

We often hear that global economic inequality is increasing (e.g., Alvaredo et al. 2017). It has been increasing for centuries and continues to do so. But do we understand why? How should we go about conceptualizing the driving forces of increasing inequality? Is there something unequal or unfair about international trade? If so, we are not likely to get the answer from mainstream neoliberal economists, who tend to be eager supporters of free trade. Their faith in free international trade can be traced to nineteenth-century economic thought in the core of the British Empire, particularly Ricardo's theory of comparative advantage and, decades later, the so-called Marginalist Revolution. The economic models developed at this time – and significantly in this *place* – focused on the mechanisms that determined market prices.

What is the most analytically robust way of conceptualizing "unequal exchange" to offer a platform for moral and political denunciation of the neoliberal world economy, without drawing on shaky, contested, or normative assumptions? In the early 1970s, I recognized the fundamental significance of a general Marxist perspective for answering this question, but precisely because of this commitment, I also recognized that we need to explore the reasons why, by and large, Marxist economic theory fails to persuade the mainstream economic profession. How are mainstream economists able to remain unperturbed by the disastrous distributive and ecological repercussions of world trade?

To make a potentially very long story short, we can recall that Marx's revolutionary analytical strategy was to penetrate the ideological veil of market prices by arguing that, even though capitalists purchased labor power at its correct exchange-value (equivalent to its costs of subsistence), such a transaction implied exploitation of workers in the sense that the use-value of labor power in production can yield exchange-values in excess of wages. As several colleagues have reminded me over the years, according to Marx, this was *not* an "unequal" exchange. In other words, his assumption was that market exchange conducted by means of monetary prices was to be viewed as "equal," but that its exploitative implications were inherent in the distinction between the exchange-value and the use-value of labor power. This was Marx's platform for denouncing the market, and it continues to be the foundation for most Marxist critiques of the globalized market economy today.

However, to say that the exchange of a worker's labor power for wages is *not* unequal is mere semantics. The capitalist obviously receives "more" from the worker than the wages with which he compensates him or her, or there would be no reason to speak of a "surplus" pocketed by the capitalist.[1] To maintain that labor power is purchased for "what it is worth" on the market is to subscribe to the illusion of market reciprocity, rather than to recognize that the very notion of "worth" or "value" has

---

[1] The Marxist economist Paul Burkett ([2005] 2009: 192), citing Marx, writes that the "apparently equal exchange of the worker's labour-power for its value thus 'turns into its opposite ... the dispossession of his labour.'" If, as Marx asserts, the capitalist only pays the worker for a part of his or her workday, this is obviously tantamount to saying that the worker's labor is "underpaid."

the ideological function of concealing unequal exchange. Apparently adhering to the mainstream assumption that market prices express a principle of reciprocal exchange, yet convinced of the exploitative logic of capital accumulation, Marx was compelled to postulate the uniquely generative capacity of labor power. As I shall show, however, this argument is problematic for those of us who wish to establish the occurrence of what I refer to as *ecologically unequal exchange*.

Applying the conventional Marxist perspective, we should conclude that the market prices of natural resources do indeed reflect what they are "worth," and – given their alleged incapacity to yield a surplus – that there can be no foundation for proposing that trade in natural resources is unfair, exploitative, or in other ways morally reprehensible. Although the Marxist analysis of capitalism has exposed the ideological capacity of mainstream economics to obscure the very *possibility* of exploitative exchange on a free market, this analysis thus excludes an account of *ecologically* exploitative exchange. This poses a major theoretical conundrum for ecological Marxists who believe that labor power is not the only resource that is systematically appropriated on the world market for purposes of capital accumulation (cf. Foster and Holleman 2014).

Drawing on Marxist theory, Arghiri Emmanuel (1972) explained how international trade generates increasing inequalities between nations by showing that there is a systematic net transfer of embodied labor time from lower- to higher-wage countries. As labor in Marxist theory is the only source of surplus value, this approach is compatible with the proposition by world-systems analysts that unequal exchange can be defined as the "constant flow of surplus-value from the producers of peripheral products to the producers of core-like products" (Wallerstein 2004: 28). However, to argue that there is also an asymmetric global transfer of natural resources that contributes to capital accumulation in the core gives us two options: either to propose that natural resources, too, can generate surplus value (cf. Bunker 1985; Odum 1988), which at first glance would appear to contradict Marxist theory, or that exploitation must be conceptualized as a general feature of globalized market exchange that is obscured by the elaborate justification, in mainstream economic models, of free trade.

Both these perspectives could be supported analytically, and both can be seen as compatible with Marx's conviction that bourgeois economic thought obscures exploitative exchange, but I believe that the latter perspective is the most conducive to a radical rethinking of unequal exchange. Rather than let market prices define what is to be viewed as

Resource	USA	EU	JAPAN
Raw material equivalents	3.7 gigatons	6.1 gigatons	2.9 gigatons
Embodied energy	10.6 exajoules	17 exajoules	4 exajoules
Embodied land	1.1 mill. sq. km	3.1 mill. sq. km	1.3 mill. sq. km
Embodied labor	96 mill. person-yrs	120 mill. person-yrs	35 mill. person-yrs

FIGURE 8 Net imports of embodied resources to world-system cores in 2007 (data compiled by C. Dorninger)

equal ratios of exchange, based on what commodities are worth, we must understand the entire discourse on a commodity's value as conditioned by an inside view of capitalism rooted in bourgeois economic thought. Like the worker's sale of his labor power for wages and the asymmetric transfers of embodied labor exposed by Emmanuel (1972), world market trade orchestrates continuous asymmetric transfers of embodied land, energy, and materials that contribute to capital accumulation in core areas of the world-system (Dorninger and Hornborg 2015; figure 8).

To recognize such asymmetries as the essence of unequal exchange and exploitation, obscured by economic representations of free market exchange, we need to be able to analytically distinguish between the biophysical aspects of world trade, on the one hand, and the economic categories through which it is represented, on the other. This approach is the only conceivable way of establishing the occurrence of exploitation, and it remains, in my view, the most essential contribution of Marxism. As I hope to show, however, it requires a critical examination of the concept of "use-value," because this concept, designed to challenge the mainstream preoccupation with exchange-values, ultimately confuses the biophysical aspects of "use" with the semiotic or cultural category of "value."

A Critique of Exploitation by Means of the Market, but without the Notion of "Use-Value"

In Chapter 4 we reviewed the approach of Karl Polanyi ([1944] 1957), who was severely critical of the inequalities generated by the market, but without resorting to Marxist value theory. What Karl Polanyi called "the idea of the self-regulating market" and the "fictitious" commodification of labor and land is equivalent to the notion that human time and natural space should be assigned monetary values and managed as commodities in production and exchange. This in turn presupposes the notion of general purpose money: an abstract metric for quantifying the ratios at which virtually anything can be exchanged on the market. Thoroughly versed in economic anthropology, Polanyi knew that this notion was peculiar to the new form of social organization and worldview that accompanied the emergence of industrial capitalism in nineteenth-century Europe. Like Marx before him, he was thus able to reflect on the transformative societal and cosmological repercussions of money and market organization.

Both Marx and Polanyi recognized that the notion of money was implicated in the exploitative social processes that were requisite to, and inherent in, industrial capitalism. However, neither of them appears to have conceptualized the familiar idea and artifact of money *itself* as the root of these exploitative processes. Marx certainly revealed how monetary exchange-values mystified the exploitation of labor, and Polanyi showed how the disembedded, unregulated market would tend to impoverish both labor and land, but neither suggested that it was the very *idea* of money that was the first thing that needed to be challenged. Each in his own way thus identified the deleterious operation of general purpose money, but finally accepted its existence insofar as they did not advocate abandoning it as a primary requisite to their suggestions on how its pernicious implications could be curtailed or domesticated by society. For Marx, a future communist society would be able to dispense with money, but only after capitalist property relations had been overthrown (Nelson 2001b). For Polanyi, the idea of the self-regulating market should be abandoned, but only in the sense that society acknowledged the necessity of protecting itself from its most injurious consequences. The extent to which Marx's and Polanyi's analytical frameworks may have been constrained by conceptual categories fundamental to modern economies – such as general purpose money and its various implications – remains to be unraveled.[2] The core component of their argument is the

---

[2] My point is that theories of the underlying logic of modern capitalism should not use the internal assumptions of that economy as conceptual tools with which to analyze its operation, not to mention that of noncapitalist economies. No less than the mainstream economic concept of "utility," "value" represents an inside view of capitalism. To say that a certain commodity has an *objective* value – or to claim that labor is the "common substance" of all value, even in ancient Greece (Marx [1867] 1976: 151–152) – is to be

recognition that monetary exchange-values conceal and distort significant aspects of whatever is being exchanged, and in doing so contribute to growing inequalities. My main question is how to conceptualize such aspects, if not in terms of monetary value.

To me, a "Marxist" perspective represents the fundamental insight that *regular market exchange conceals asymmetric transfers of resources that contribute to the accumulation of capital in the hands of some individuals and groups, at the expense of other market actors.* Others may consider other aspects of Marxist theory more central, but this is what I take to be the most significant core of a Marxist position. My aim is to show how this general perspective can be extended beyond the exploitation inherent in wage labor and the international trade in embodied labor power (Emmanuel 1972) to the trade in several other kinds of resources as well (Hornborg 2001, 2013a, 2016a; Dorninger and Hornborg 2015). To establish the existence of asymmetric exchange, mystified by the apparent reciprocity of market prices, we need to rigorously conceptualize the precise substance of the so-called use-values that are being appropriated in production and world trade. I concur with several proponents of an ecological Marxism in suggesting that this implies acknowledging the appropriation of other kinds of biophysical, potentially value-enhancing inputs in production, in addition to the energy of labor power. However, rather than proposing an extended concept of use-value, I will argue that the concept is intrinsically misleading. This conclusion draws on long-standing efforts to understand the global ecological and economic crises of our times, viewed through the concerns with uneven accumulation and exploitation that Marx so incisively pioneered.[3] What I take to be the core idea of Marx's analysis – the acknowledgment of a materially asymmetric exchange that is obscured by flows of exchange-values – can be usefully expanded and applied to modern discussions about sustainability and global justice.

---

    constrained by the categories of our own society in our attempts to develop a comparative understanding of human economies. In this sense, it is as ethnocentric as it would be for a citizen of ancient Athens to explain slavery by referring to how the characteristics of certain people make them suitable for serving as slaves.

[3] In an early attempt (Hornborg 1992), I too suggested that the productive potential that was transferred to centers of capital accumulation should be conceptualized as underpaid use-value, but given the kinds of analytical problems with this approach, I later decided to clearly distinguish the physical from the semiotic.

The history of Marxist theory can be read as a struggle to analytically integrate the social sphere of exchange-values and the natural sphere of biophysical metabolism. The theoretical framework developed by Karl Marx and Friedrich Engels in the mid-nineteenth century to explain the underlying logic of industrial capitalism was founded on the insight that the economy can only be understood by juxtaposing features of society and nature. Indeed, this is arguably the essence of historical materialism as a scientific theory of capital accumulation. A challenge for Marx was to account for the conversion, in production, of labor energy into exchange-value.[4] As a century and a half of debate illustrates, however, his conceptualization of this relation cannot be viewed as final. The modern concept of energy was not established in physics until the 1860s, and to the extent that early economists aspired to model their science on physics, they were still unable to draw on the Second Law of Thermodynamics, the so-called entropy law (Georgescu-Roegen 1971; Mirowski 1988). A long line of theorists have over the course of a century and a half aspired to reinterpret or modify Marx's framework. While united by the conviction that his critical understanding of industrial capitalism (as based on compulsive accumulation, exploitation, and alienation) is fundamentally valid, their debates have highlighted some inconsistencies and ambiguities in his texts. A recurrent problem for these theorists has been how to handle the relation between the economic and the biophysical – the social and the natural. Over the past 30 years, augmented by the expansion of environmental concerns, this interdisciplinary dilemma has become the focus of debates within the field known as *ecological* Marxism (cf. O'Connor 1998; Burkett [1999] 2014; Foster 2000). Further on in this chapter, I shall briefly review the history of such debates. I will suggest that the concept of use-value tends to constrain ecological Marxism from pursuing its concern with asymmetric social transfers of material resources such as labor power and other forms of energy. Finally, I shall argue that this central Marxist concern necessarily implies a reconceptualization of the phenomenon of modern technology, which remains a contested and insufficiently theorized aspect of capitalism since the Industrial Revolution. But first I will explain why I believe that the concept of "use-value" is not useful for establishing the occurrence of exploitation and unequal exchange.

---

[4] In their efforts to demonstrate that Marx was well versed in thermodynamics, Burkett and Foster (2006: 120) approvingly quote a translation of Volume I of *Capital* in which he asserts that "labour-power itself is energy" (Marx [1867] 1967: 215).

Three Approaches to "Use-Value"

Steve Keen (1993a, 1993b, [2001] 2011) has argued for a rethinking of the role of use-value and labor power in Marxist value theory, particularly following the publication of *Grundrisse* (Marx [1857] 1973). He notes that Marx "used the *quantitative* difference between the exchange-value of labor-power and its use-value to uncover the source of surplus value in the transaction between worker and capitalist" (Keen 1993a: 112; emphasis in original). Indeed, as I have noted previously, the only reason to speak of a "surplus" is that one quantity (exchange-value) is subtracted from another (value-creating potential). From this perspective, use-value and exchange-value appear to be quantifiable and comparable measures, and the common metric for comparing them is economic value (money), which in the Marxist model corresponds to abstract labor time. According to Marx, the peculiarity of the commodity of labor power is that the daily cost of maintaining it (its exchange-value) is lower than its potential contribution to the value of production (its use-value). Although the capitalist purchases labor power at its market value, in employing its use-value in production, he ends up with more value than he has invested. But if Marx had "properly applied his own logic," Keen claims, he would have conceded that labor cannot be the only source of surplus value. Citing Marx's own argument, Keen argues that labor power is not the only input in production that can have a use-value greater than its exchange-value, and that could thus contribute more to the value of a commodity than its own exchange-value, that is, be a source of surplus value.

According to Keen, in conflating the physical deterioration of a machine's productive – that is, value-creating – capacity and its monetary depreciation, Marx inadvertently concurred with neoclassical economics in equating usefulness (use-value) and cost (exchange-value) in a notion of abstract utility (Keen [2001] 2011: 436–438). His version of the labor theory of value, Keen argues, is founded on the inconsistent assertion that the contribution of machinery and raw materials to value production is not based on their use-value, as Marx argues is the case with labor power, but exclusively on their exchange-value, which means that they can transfer no surplus value to the product. However, Keen refers to Marx's conclusion in *Grundrisse* that the use-value of the machine is "significantly greater than its value" (Marx [1857] 1973, cited in Keen [2001] 2011: 438–439). Contrary to Marx, Keen ([2001] 2011: 441–442) asserts that "surplus value – and hence profit – can be generated from any input to production."

A second approach to use-value is that of Paul Burkett ([1999] 2014) and John Bellamy Foster (2014), who equate the concept with the "natural-material" aspects of a commodity. Quoting Marx ([1857] 1973: 284), Burkett ([1999] 2014: 206) exemplifies the use-value aspect of commodities to a worker by their capacity to serve as "a means of subsistence, objects for the preservation of his life, the satisfaction of his needs in general, physical, social etc." Marx's inclusion of the worker's "social" needs in the definition of use-value is fundamental to the third interpretation to be discussed here; suffice it to say that this definition extends far beyond the concept of "natural-material" subsistence and physical needs emphasized by Burkett and Foster. However, Burkett's and Foster's definition of use-values as "natural-material" is essential to their conviction that an ecological perspective is inherent in Marxism (Foster 2000; Burkett [1999] 2014). Burkett argues that the use-value of labor power is its physical dimension, to be contrasted with its monetary exchange-value. From this perspective, the capitalist appropriation of surplus value from workers and their families is not simply a matter of making profits from keeping monetary costs lower than incomes, but a "material, biological, and energetic extraction process" (ibid., xviii). Burkett and Foster have thus described the worker's sale of his labor time as "an energy subsidy for the capitalist" (Foster and Burkett 2008: 26) and explicitly refer to Marx's "energy income and expenditure approach to surplus value" (Burkett and Foster 2006: 126).

If the extraction of surplus value is a matter of harnessing the physical energy of labor power, as Burkett proposes, it would be invalid to propose that the use-value of labor – as opposed to its exchange-value – can be quantified in money. If use-value is the physical quality of com-modified labor that makes a material process of production possible, it must be conceptualized in material rather than pecuniary terms. It refers to metabolic aspects, which means that it can only be quantified in terms of physical metrics such as energy. Moreover, this material capacity to contribute surplus value to a production process cannot be reserved for labor power alone. If labor power is indeed energy, as Marx asserted, there is no other reason to conclude that this particular kind of energy alone has a unique value-creating capacity than that Marx says so, echoing a widespread assumption in his day. While it is essential to acknowledge, as Marxism does, that there are other kinds of transfers going on than the monetary flows that preoccupy neoclassical economics, such invisible, asymmetric transfers include embodied land, energy, and materials as well as labor. The Marxian resistance to neoclassical economics can thus be extended to an acknowledgment of uneven flows of other biophysical resources, in addition to labor (cf. Foster and Holleman 2014).

To assert that the exchange-value of labor power is lower than its contribution to the value of the product is indistinguishable from making similar propositions about other energy inputs such as from draft animals or fuels. A horse can also work more hours than the time it takes for it to pay its keep. The inputs of resources such as coal, iron, and cotton fiber in nineteenth-century British textile production were as indispensable to capitalist profits as industrial labor, and the asymmetric transfers of such resources – gauged in terms of embodied labor and land – to the textile-producing districts were no less crucial to the accumulation of industrial capital. Understood in physical terms, as suggested by Burkett and Foster, there is no essential difference between the labor power of industrial workers and the muscular energy of slaves or horses.

A third interpretation, finally, follows Jean Baudrillard ([1972] 1981) and Marshall Sahlins (1976) in observing that use-values are always mediated by culture. In other words, they are symbolically constituted. There is no purely "natural-material" aspect of a commodity for which a worker is willing to exchange his or her labor time. Not only are even the foodstuffs that are understood to provide the worker's subsistence and cater to his or her physical needs largely culturally selected, but much of a modern worker's wage will be spent on commodities that satisfy social needs that seem quite arbitrary and have very little to do with the "natural-material" preservation of his or her life. As modern sociological studies of consumption have abundantly shown (e.g., Lash and Urry 1994), the use-values of market commodities tend to reflect their cultural-symbolic rather than natural-material aspects.

The tension between these two conceivable meanings of use-value can be traced to Aristotle's original formulation in *Politics* in the fourth century BC:

Every article or property has a double use; both uses are uses of the thing itself, but they are not similar uses; for one is the proper use of the article in question, the other is not. For example, a shoe may be used either to put on your foot or to offer in exchange. (Aristotle 1962: 41)

Aristotle's example does not only distinguish between use-value and exchange-value; it also evokes two possible dimensions of a commodity's use-value. The notion of "proper use" here can be subdivided into two aspects: it can signify both a person's subjective and culturally constituted need of a shoe and the objective, material capacity of a shoe to satisfy that need. Baudrillard and Sahlins join neoclassical economics in focusing on the former, that is, "utility" as a *relational* quality defined by the subjective desires of the consumer, whereas Burkett and Foster emphasize the latter. In response to Burkett, Baudrillard would have objected that the objective capacity of a commodity to satisfy a specific need cannot simply be reduced to its natural-material aspects, as if it were an inherent physical attribute of the commodity, irrespective of the consumer. Regardless of the analytical disjunction between the subjective and objective aspects of use-value, it is evident that neither of these aspects can be quantified in money. If conceived as a physical property of the commodity, use-value must be measured in other than monetary terms, and if conceived as a semiotic quality, it cannot be quantified in any terms at all.

What, then, *is* "use-value"? Whenever I have pointed out that use-value must be understood as to some extent symbolically constituted, Marxists have assured me that this is quite compatible with Marx's position, but if the use-value of a commodity simultaneously implicates its objective, natural-material aspects, as in the interpretation of the use-value of labor power as energy, the concept collapses into amorphousness. Marx in the first chapter of *Capital* appears to include in "use-value" anything that someone might be willing to spend money on, from Bibles to clothing (Marx [1867] 1976),[5] which leads me to wonder how it could be distinguished from the neoclassical economist's notion of "utility." As utility in mainstream economic theory is equivalent to exchange-value, however, such a definition would contradict the Marxist framework.

Given that the separate theoretical perspectives pursued by Keen, Burkett, and Baudrillard all have their validity, the tension between their approaches reveals the analytical limitations of a concept that simultaneously lends itself to interpretations emphasizing monetary quantification, asymmetric transfers of energy, and cultural semiotics. Unsurprisingly, this ambiguity has generated considerable confusion. At stake are, among other things, the capacity of the concept of use-value to undergird a labor theory of value, and the argument that natural resources are underpaid in

---

[5] On the very first page of *Capital*, Marx ([1867] 1976: 125) asserts that it makes no difference whether the needs satisfied by a commodity arise "from the stomach, or the imagination."

world trade.[6] Rather than circumambulate these analytical problems, an interdisciplinary theory of exploitation and unequal exchange will need to unravel the conundrum posed by the ambiguous notion of use-value. As the next section will illustrate, the history of debates in ecological Marxism largely reflects the conceptual constraints posed by such ambiguities. Although a powerful objection to mainstream economic thought, the proposal of a surplus-generating use-value unique to labor power is not the only possible analytical strategy through which the mainstream apotheosis of market prices can be confronted. In the remainder of this chapter, I will attempt to show how the concerns of ecological Marxism can lead to a renewed challenge to the illusion of market reciprocity. Contrary to so-called Promethean strains in the Marxist tradition, such a challenge must also be aimed at the ontology of technological progress.

A Brief History of Ecological Marxism

In the 1980s, after two decades of environmentalist discourse, several socialist theorists criticized mainstream Marxism for its neglect of ecological issues. A common denominator of this critique was the conviction that Marxist economics did not sufficiently recognize the materiality of the natural environment. Some focused on how ecological conditions posed physical constraints on economic activity, others argued for a more consistent integration of natural science in the theoretical framework of historical materialism, including the acknowledgment that nature, as well as labor, can be a source of economic value. An influential example of the latter was the observation by Joan Martinez-Alier and José M. Naredo (1982) that the response, in the early 1880s, of Friedrich Engels to the ideas of Sergei Podolinsky was symptomatic of the difficulties of integrating Marxian value theory and thermodynamics. The observation is repeated in Martinez-Alier's (1987) book *Ecological Economics*.

Podolinsky had recognized that Marx was aiming to articulate a natural science of society by explaining economic value with reference

---

[6] Most current advocates of a theory of ecologically unequal exchange define it in terms of asymmetric flows of material resources, rather than in terms of transfers of surplus value (Jorgenson and Clark 2009; Hornborg and Martinez-Alier 2016; Frey, Gellert, and Dahms 2017, 2018). While capitalists will obviously focus on the augmentation of surplus monetary value, this inside category of capitalism must be distinguished from the material metrics – embodied labor and other resources – by which the objectively asymmetric exchanges that make their profits possible can be gauged.

to the investment of physical labor power in production. Engels nevertheless dismissed his proposal that Marx's concept of surplus value could
be understood in terms of energy. In hindsight, this failure of communication illuminates how Marx's aspiration to analytically merge nature
and society was blocked by the incommensurable ontologies of physicists
and economists, and by the conceptual confusion accompanying the
historical transition from organic to inorganic energy in production.
Foster's and Burkett's efforts (Foster and Burkett 2004, 2008; Burkett
and Foster 2006) to justify Engels's dismissal of Podolinsky's energy
theory of value as reductionist are valid, but unfortunately make no
effort to understand what drove Podolinsky to contact Marx in the
first place, or how his proposal ultimately highlights a fundamental
ambiguity in the labor theory of value. Paradoxically, Foster and Burkett
repeatedly assert that Marx's theory of surplus value is a matter of
energy flows between workers and capitalists, while apparently adhering
to the core Marxian conviction that the increase of economic value in
production can be analytically derived from expenditures of labor power.
A step forward in transcending the ontological divide revealed by this
conundrum would have been to conclude that *any* energy theory of
economic value is reductionist (Georgescu-Roegen 1979), regardless of
whether that energy is provided by fuels, draft animals, *or human labor*.
Energy and economic value must be kept analytically distinct. To the
contrary, however, Foster has endorsed the explicit energy theory of
value of Podolinsky's intellectual heir Howard T. Odum (Foster
and Holleman 2014; cf. Odum 1971, 1988). Such continued intuitions
about the convergence of energy and labor theories of value (cf.
Lonergan 1988) highlight the problematic role of the concept of use-
value in ecological Marxism. They reveal how a theory that claims to
analytically derive exchange-value from inputs of labor power, defined
in biophysical terms, is as flawed as the energy theory of value proposed
by Podolinsky.

Ted Benton (1989) identified a general technological optimism in
Marxist theory that underplayed natural limits to production and thus
seemed to make it ill-equipped to deal with environmental problems. He
traced this "Promethean" approach to the discourse of classical political
economy in which Marx was involved. This historical context contributed to his labor theory of value as well as his Enlightenment belief in the
progress of technology. Similar, ecological critiques of Marxism were
published by André Gorz (1994), Alain Lipietz (2000), and Joel Kovel
(2002). Several ecological feminists have also questioned mainstream

Marxian productivism, as well as the labor theory of value (Soper 1991; Salleh 1994; Brennan 1997).[7]

In 1988 James O'Connor launched the first issue of the eco-Marxist journal *Capitalism Nature Socialism* with a theoretical introduction (O'Connor 1988) outlining his proposition that capitalism suffers from not one but two crisis-generating contradictions. In addition to the tendency of capital to keep wage-based demand for its commodities below what is necessary to realize their surplus value, the second contradiction of capitalism denotes the tendency of capital to degrade its "conditions of production," including the natural environment. Ten years later, the text was reprinted in O'Connor's (1998) collection *Natural Causes: Essays in Ecological Marxism*, which elaborates and exemplifies the role of ecological conditions in capitalist crises. The aim of O'Connor's work was to complement the classical Marxist framework by developing its potential for theorizing natural constraints and identifying the actors of socio-environmental conflicts, demonstrating that the global protagonists of environmental justice are generally external to the industrial proletariat. The two contradictions of capitalism, in this view, thus oppose different categories of social actors. As the global environmental justice movement makes clear, industrial workers are not the only victims of capitalist exploitation, and labor power is not the only productive resource subject to appropriation.

In a review of O'Connor's book, Burkett (1999) criticizes him for accusing Marx and Engels of having slighted ecological concerns, rather than acknowledging the ecological insights of the nineteenth-century founders of Marxism. In a book published the same year, Burkett ([1999] 2014) uses exegesis of nineteenth-century texts to demonstrate that Marx's theory of value was indeed intended to accommodate the natural, material aspects of the capitalist economy by conceptualizing them as "use-values" obscured by the exchange-values of the market. The following year, Foster published his book *Marx's Ecology*

---

[7] Whereas a common defense of Marx's labor theory of value is that it represents an account of the specific operation of capitalism, rather than of his own convictions, this is contradicted by Marx's ([1867] 1976: 151–152) assertion that Aristotle was unable to *see* the "common substance" of all value as labor because ancient Greek society was founded on slavery rather than capitalist commodity production. If, according to Marx, all value derives from labor even in ancient Greece, his labor theory of value cannot merely be an account of the operation of capitalism. Conversely, if Marx's concept of "value" derives from and refers to specific features of capitalist society, it should be invalid to apply it to the analysis of ancient Greek economy.

(Foster 2000), in which he similarly showed that there is plenty of textual evidence that Marx was highly concerned about ecological issues, particularly the "metabolic rift" in nutrient flows between urban and rural areas. The identification of a metabolic rift in the material flows of industrial capitalism has been extended from concerns with soil nutrients to wider asymmetries in the global metabolism of modern world society (Foster, Clark, and York 2010).[8] In Foster's view, the abandonment of Marx's ecological concerns in twentieth-century Western Marxism was a consequence of the critique of technological rationality and Enlightenment by the Frankfurt School (Foster and Clark 2016).

In several publications since then, Burkett and Foster have continued to defend Marx and Engels against the charges of Prometheanism and ecological ignorance that have been directed at the founders of Marxism by environmentalists and ecological socialists (Burkett [2005] 2009; Foster et al. 2010; Foster and Burkett 2016). This defensive posture has puzzled some ecosocialists, who believe that more would be gained from updating and modifying Marx's theoretical framework than from denying any indication of inconsistency or ambiguity in it. Foster and Burkett have shown that the foundations for an ecological socialism can indeed be unearthed in the original texts of Marx and Engels, but this does not neutralize their contradictions, or the next to universally Promethean interpretations and policies that they have generated. As one reviewer (Rudy 2001) concludes, Foster's demonstration that Marx's concerns were genuinely interdisciplinary lays the groundwork for an ecological Marxism that Marx had been unable to develop.

The pervasive dilemma of integrating knowledge of the operation of society with knowledge of the operation of nature has generated recurrent problems for Marxist conceptualizations of nature. Modern proponents of what Foster and Clark (2016) call Western Marxism (e.g., Smith 1984; Harvey 1996; Castree 2014) have downplayed the significance of nature as an autonomous and constraining reality beyond its construction by society. For some of them, the very urge to distinguish between society and nature is to succumb to a misguided dualism.

The inclination to challenge such dualism has led Jason W. Moore (2015) to ally himself with the deliberations of so-called posthumanism, and to criticize Foster from that obscure vantage point (see Chapter 11).

---

[8] The concept of metabolic rift is thus compatible – perhaps even synonymous – with the concept of ecologically unequal exchange, but I would phrase it in terms of an asymmetric transfer of resources rather than "values."

However, like Latour and other posthumanists, Moore confuses dualism and binary analytical distinction. While Moore is to be credited for his earlier work on the history of commodity frontiers in the world-system (e.g., Moore 2007), his "world-ecology" program thus does not contribute a clear analysis of the relation between the social and the natural. Although his concept of "work/energy" indicates that he is aware that the labor theory of value needs to be transcended in favor of a more general concern with the appropriation of energy, he does not provide a coherent critique and reconstitution of ecological Marxist theory. To collapse the natural and the social into an amorphously unified whole does not contribute to the analytical aspirations of historical materialism.

Moore's emphasis on capitalist strategies to increase labor productivity by appropriating "cheap" resources from elsewhere does not lead him to reconceptualize the very ontology of technology as a global societal phenomenon (cf. Hornborg 2001). Moreover, to say that some of the "work/energy" appropriated by capitalists is "unpaid" is indistinguishable from saying that "work/energy" as a whole is "underpaid," once again illustrating how the market shapes the conceptual horizons even of its critics. It thus also echoes mainstream concerns with internalizing environmental "externalities" such as the cost of "ecosystem services" (Moore 2015: 64). Whereas Moore at one point rejects the idea that capital could possibly pay the "true costs" of resource use (ibid., 145), this does not stop him from constantly phrasing its appropriation of resources in terms of "cheap" or "unpaid" nature. Such conceptualizations of asymmetric exchange are highly misleading. Notions of costs and underpayment implicitly refer to assumptions about monetary values, but without questioning the artifact of money. In subtle ways money thus continues to constrain our thinking about exploitation and nature, even in ecological Marxism.

As Marx demonstrated, capitalism is a process of accumulation, but to conceptualize it in terms of access to "cheap" resources is to be deluded by the monetary idiom that serves to veil its biophysical asymmetries. There can be no analytically rigorous definition of "cheap" because it gives the impression that the crux of the matter is the relation between a commodity's objective properties (its "value") and its price. Appropriation can be identified only by establishing net flows of resources. Prices certainly orchestrate such flows, but the existence of those flows must be documented using biophysical – rather than monetary – measures. To say that something is "cheap" is to assert that its price does not match its value, but as Ole Bjerg (2014: 19–45) has shown, using Slavoj Zizek's categories "the symbolic" and "the real," the notion of a correct price is founded on the ultimately illusory idea of an objective value. Bjerg's analysis is concerned with the evaluation of stocks, but it is equally applicable to commodities. The urge to imagine value as an objective property attributable to natural phenomena, rather than a transient and subjective relation contingent on social context, has spawned both labor and energy theories of value. It ultimately raises the question whether value belongs to nature or society, to which the incontrovertible answer must be the latter. This conundrum continues to hamper a clear grasp of what "ecologically unequal exchange" really signifies (Gellert 2019). As I argued more than twenty years ago (Hornborg 1998a), because production of economic value is inexorably destructive of what we might call (productive) *potential* (or "negative entropy," see Georgescu-Roegen 1971), it is contradictory to equate such potential (e.g., the availability of energy) with value, as have Bunker (1985), Odum (1988), and Foster and Holleman (2015). While it is not sustainable for human societies to generalize the attribution of higher values to things the more destruction they represent, it is no more justifiable to attribute higher values to them the less useful they are to humans. The only way to curb the globalized logic of the former without proposing the latter is by spatially constraining the commensurability of commodities.

If there is no way of establishing or measuring a commodity's non-monetary value, the notion of "cheapness" is analytically useless. It can only mean that a commodity is relatively less expensive than other alternatives on the market. Assessing a commodity's price should not be conceived as a judgement about the extent to which its price corresponds to its objective properties, but about how its price relates to that of other commodities on the market, particularly those commodities into which it is transformed. This relation is not usefully illuminated by asserting the "cheapness" of inputs in production but can be examined empirically by quantifying the net flows of resources that result from discrepancies in market prices, that is, ecologically unequal exchange (Dorninger and Hornborg 2015).

Money as a Signifier without a Signified

Marx rarely used the word *capitalism*. He analyzed the operation of *capital*. What, then, is the relation between his concept of capital and general purpose money? This question has been posed and answered in different ways over the past century and a half. David Harvey (2018: 5) explains that, in Marx's view, money is the "representation of value." Value, in turn, is "socially necessary labour time," which Marx visualized as an "immaterial but objective" relation, of which money is the material representation (ibid.). But money is a "distorted" representation of value (ibid., 51), and the various discrepancies between money and the value that it is "supposed to represent" (ibid., 69, 94, 106) account for the recurrent crises of capitalist society. Harvey (ibid., 172) observes that "[d]aily life is held hostage to the madness of money."

Money is unquestionably a representation, that is, a *sign*, and it is indeed legitimate to expect a representation or sign to signify (i.e., stand for) something. Such an expectation is fundamental to all semiotic inquiry. It can be argued, however, that the money sign is uniquely devoid of a specifiable referent, that is, that it does not signify anything beyond itself. The peculiarity of money is its capacity to assume any meaning that its owner bestows upon it (Hornborg 2016a: 39–40). Rather than propose that money relates to value as signifier to signified, we might thus turn the relation around and suggest that *value is a concept that refers to money.* The very idea that goods and services exchanged in a market have an objective value, implicating some common standard, is derived from the artifact of money.

The notion that value is an objective property of social relations that is imperfectly represented by money was Marx's way of deconstructing and challenging mainstream understandings of free market capitalism as benevolent and fair. It made it possible for him to theorize exploitation. But it is not the only possible analytical strategy for explaining how free market exchange generates inequalities by systematically mystifying them. It is possible to argue that market exchange orchestrates asymmetric flows of material resources (including labor power), which unevenly contribute to the technological capacity of different groups to produce exchange-values, without positing that those material resources have an immaterial value that exceeds their monetary price. From this perspective, the artifact of money does not "represent" the kinds of social relations that we attribute to capital, or capitalism – it *generates* them. Rather than mystifying social relations conceptualized in terms of value, money *is* the essential relation of production (cf. Harvey 2018: 59).

To propose that the logic of capital is the consequence of social relations that are more fundamental than those generated by general purpose money is similar to suggesting that the trajectory of a game of Monopoly is not a product of the rules of the game, but of the social relations among the players. In the "insanely speculative economy" (ibid., 60) of recent decades, however, it has become increasingly plausible that the logic of money is the motor of capitalism. In this situation, writes Harvey (ibid., 104), it is tempting to see money as "the essence of what capital is about and to redefine capital as money in motion rather than value in motion." But to focus on money and market prices and "erase the relation between prices and values," he asserts (ibid., 63), would be to discard a "standpoint from which to mount a critique of the monetary representations." According to Harvey (ibid., 105), Marxists proposing a monetary theory of value (e.g., Moseley 2015) might join the neoclassical economists in taking "the reassuring and comforting position that all will be well with global capitalism once it has settled back into the equilibrium conditions dictated by perfectly functioning and properly regulated price-fixing markets." If we accept a "purely monetary theory of capital," Harvey continues (2018: 106), we lose the capacity to critique the "increasing concentration of monetary wealth ... at the expense of the rest of humanity."

The bottom line of this argument is that to equate capital with general purpose money would be to renounce the critical approach that animates the Marxian project. It would be to subscribe to the mainstream neoclassical idealization of the market, and to abandon the unique conceptual lever ("value") that enabled Marx to deconstruct and challenge its inequalities. But a theory of unequal exchange does not need to assume "immaterial but objective" values that are misrepresented by money prices. It can instead expand on Marx's crucial insight that money prices serve as a veil that mystifies asymmetric material transfers between market actors. Such an approach is diametrically opposed to neoclassical economics, for which other measures of exchange than money – for instance, embodied labor, land, energy, or materials – are simply irrelevant. Although the latter metrics should not be confused with values, they share with the Marxian concept of value a focus on the significance of nonmonetary aspects of the economy. Contrary to Harvey's conviction, to acknowledge that money is the root of all evil is not necessarily tantamount to accepting it as inevitable.

We may have to concede that mainstream neoclassical economists are right in that the various phenomena they study – however alarming or "insane" (Harvey 2018) they are – can be explained in terms of the logic of money, without positing some metaphysical, immaterial notion of value. The "madness of economic reason" (ibid.) lies in not acknowledging that this logic is so destructive of the conditions of human life that it must be fundamentally redesigned. To delegate the organization of world society to the rational management of general purpose money is to give market actors automatic incentives to promote the most "efficient" production processes, that is, those based on a minimum of labor costs and environmental considerations. The aggregate consequences of the logic of money – whether pursued by corporations, nations, or individual consumers – thus propels globalized capitalism toward increasingly obscene injustices and ecological degradation. Rather than simply model the myriad details of this trajectory, as if observing the inexorable implications of natural law – or the operation of a "machine" suggesting "the epitome of perfection" (Harvey 2018: 174) – it should be incumbent on the discipline of economics to question its foundational assumptions regarding the benevolence of general purpose money and free market trade. A theory of ecologically unequal exchange poses no less of a challenge to mainstream economics than does classical Marxism. However, it does so by referring not to a contested notion of immaterial value, but to incontrovertible flows of material resources. Moreover, it does not define the fundamental problem in terms of an abstract mode of production, which requires detailed and voluminous scrutiny, but in terms of the insidious implications of the very artifact of general purpose money.

An Ecological Approach to Technology and Unequal Exchange

In retrieving the original intentions of Marx and Engels, Foster and Burkett have provided the foundations for a renewed effort at integrating our understanding of the social and natural dimensions of industrial capitalism. But to pursue such a genuinely interdisciplinary understanding, we must unpack the ambiguous concept of underpaid use-values. As illustrated by the contrast between the perspectives of Burkett and Baudrillard, in conflating a commodity's material properties and its anthropocentric (and inevitably cultural or semiotic) *value*, it misleadingly suggests that there is a common metric for assessing the discrepancy between its natural and social features. The implicit notion of underpayment derives from the market, illustrating yet another conceptual constraint in Marxist thought inherited from bourgeois political economy.[9] When Foster sets out to reframe the theory of ecologically unequal exchange in classical Marxist terms, he and Hannah Holleman (Foster and Holleman 2014), like Stephen Bunker (1985) and Howard Odum (1988) before them, thus suggest that

---

[9] As mentioned, other examples include the labor theory of value and the faith in technological progress (cf. Benton 1989).

	Embodied biophysical resources (nature)		Value		Monetary market price (societal evaluation)
Neoclassical economics	(Irrelevant)		Value (utility)	▬▬	Price
Marxist economics	Embodied labor	▬▬	Value	⇌	Price
Ecological economics	Embodied energy	▬▬	Value	⇌	Price
Ecologically unequal exchange	Embodied resources	⇌			Price

FIGURE 9 The role of the concept of "value" in different schools of thought, as based on objective metrics ("nature") or subjective perceptions ("society")

the asymmetric transfer of biophysical resources on the world market is to be understood as an unequal exchange of *values*. This is to confuse a cultural and a material perspective on the economy. The significance of asymmetric transfers of material resources such as energy is not that they represent underpaid values, but that they contribute to the physical expansion of productive infrastructure at the receiving end. As I have previously emphasized, the accumulation of such technological infrastructure may yield an expanding *output* of economic value, but this is not equivalent to saying that the resources that are embodied in infrastructure have an objective value that exceeds their price. In positing a distinction between a commodity's objective value and its price, both Marxist and ecological economics suggest a framework for asserting that it may be "underpaid," but the theory of ecologically unequal exchange does not require a notion of objective value to show that market prices may orchestrate objectively asymmetric transfers of biophysical resources (Figure 9), which enable the uneven accumulation of productive infrastructure.

A reconceptualization of ecologically unequal exchange along these lines is necessary to understand the development of industrial technology – the very source of environmental degradation and charges of Prometheanism – as capital accumulation involving asymmetric transfers of biophysical resources. Labor energy is one such resource, but not the only one. It seems paradoxical that labor should be identified by Ricardo and Marx as the exclusive source of profit at the very time in history when its role in the production process was becoming decisively superseded by other sources of energy – fossil fuels – and substantially complemented with international transfers of other indispensable resources such as embodied land and materials. I believe that the explanation of this paradox is that the attribution to labor of value creation has a very long genealogy in moral philosophy and economic thought, and that it was merely being asserted in new and more elaborate ways in the intensive discourse on political economy in the nineteenth century (cf. Berg 1980). Throughout history, labor theories of value were generally connected with a widespread disdain for "unproductive" money-making such as lending money at interest (Le Goff 2012). Marx thus expressed sentiments going back to Thomas Aquinas, St. Paul, and Aristotle (Tawney 1972).

The essential role of fossil energy in defining the operation and environmental history of industrial capitalism has been reconceptualized by several researchers (Deléage 1989; Altvater 1994, 2007; Hornborg 2001, 2013b; Huber 2008, 2013; Barca 2011; Mitchell 2011; Malm 2016). Rethinking the relation between energy and exchange-value, and its transformation by the turn to fossil fuels, makes it possible to pose new questions regarding the intuitions underlying Marx's conceptual framework. It seems possible, for example, that the formal congruity between Marx's argument about the falling rate of capitalist profit and current concerns with diminishing net energy, or EROEI (Hall and Klitgaard 2011), *both attributed to increasing mechanization*, is not a coincidence. This can prove particularly significant, given Keen's ([2001] 2011: 439) conclusion that, "with the labor theory of value gone, so too would be the tendency of the rate of profit to fall, and with it the inevitability of socialism." If the falling rate of profit can be understood more generally in terms of diminishing returns on mechanization, rather than merely in terms of a lower proportion of living labor to machinery, Marx's predictions about the destiny of capitalism would not be predicated on the labor theory of value. Paradoxically, the modern relevance of Marxist theory would thus be enhanced by a critical reconsideration of its labor theory of value, and by contemporary concerns about resource depletion and environmental degradation.

The ecological Marxism developed by Foster and Burkett, building on aspirations traced to Marx's historical materialism, attempts to integrate the social and the natural within a common theoretical framework, while acknowledging *in principle* the analytical autonomy of social and natural processes. Although promising, this approach would progress further if it entailed a sympathetic critique of some of Marx's own categories, aiming to extend and expand his insights rather than merely reiterate them. As their understanding of "use-value" does fail to acknowledge the analytical autonomy of the social and the natural, Foster's and Burkett's paradoxical deliberations on the role of energy in historical materialism would be the obvious muddle with which to begin such a progressive phase of reconstitution, continuing where Marx left off.

Drawing on some of the analytical interventions reviewed here, it is possible to assemble a renewed approach to the contested arguments on appropriation and accumulation that were so central to Marx's concerns. A coherent ecological Marxism must acknowledge that other productive inputs than labor can be sources of capitalist surplus. It would recognize that the productive inputs conceptualized by Marx as use-values appropriated by capitalists comprise a wider category of biophysical resources including embodied energy, labor, land, and materials. The concept of capital accumulation, from this perspective, should imply a reconceptualization of industrial technology as a local augmentation of labor productivity through net imports of resources from elsewhere. Technologies, in other words, are not just manifestations of scientific discoveries and congealed local labor, but products of asymmetric – and incontrovertibly *social* – global resource transfers. Although it is highly pertinent to demonstrate that new technologies have locally served to increase rates of profit (Marx 1867; cf. O'Connor 1998: 201), as effective "tools of empire" (Headrick 1981), as means of controlling labor (Malm 2016), as instruments of exclusion (Winner 1980), and more generally as modes of manipulating human behavior, such insights thus do not exhaust the extent to which globalized technologies are *societal* phenomena. Beyond such tangible and conscious incentives recognized by the relevant actors at the local level, social theory must seek to identify the structural requisites of industrialization – the global conjunctures and circumstances that were not amenable to local reflection but made mechanization possible.

Capital accumulation in nineteenth-century Britain was not *only* a product of the surplus value generated by British workers but also of the colonial appropriation of resources. Nor is it meaningful to assert that technologically advanced sectors of the world-system have increased the rate of exploitation of its own workers, while neglecting that these sectors have been able to do so by simultaneously increasing the rate of net imports of resources from elsewhere. This means that modern industrial workers employed in core countries should not be regarded as the main antagonists of capitalism, but that in material and political terms they may rather find themselves its *allies*. The transfer of jobs to lower-wage sectors of the world economy in recent decades have highlighted the extent to which it is in the interest of workers in former cores to remain at work and to maintain their globally privileged levels of consumption. The greatest single challenge for an ecological Marxism is to reconceptualize the advance of modern technology, traditionally celebrated by Marxism as the progress of the "productive forces." Following Marx, Burkett's ([1999] 2014: 66) formulation that technological development employs forces "present in nature" suggests that technology is fundamentally a revelation of possibilities inherent in nature, rather than a thoroughly *societal* phenomenon contingent on specific relations of asymmetric exchange. Burkett quotes Marx's reference to the natural resources incorporated in technology as "a free gift of Nature to capital, that is, as a free gift of Nature's productive power to labour" (ibid., 71),[10] which does not acknowledge that the enhancement of labor productivity through modern technology – and thus, in Marx's view, the intensified exploitation of the factory worker – is feasible only through an increasingly asymmetric social exchange of biophysical resources on the world market.

In the same vein, Harvey (2018: 107–126) has written a full chapter on "The Question of Technology" – even referring to "technological fetishism" – in Marxist thought, without once indicating that technological progress may imply appropriation of labor and resources from other areas. The assumption seems to be that the development of the so-called productive forces during the Industrial Revolution in Britain was exclusively about "locating an energy source outside and beyond that of the manual strength of the labourer" and about "[scientifically dissecting] nature in order to exercise control" (ibid., 117–118), as if Britain's position in the eighteenth-century world-system was not implicated in the operation of industrial technology. The machine is conceived of as an innocent revelation of nature, rather than a *socionatural* machination. Ultimately, Harvey's delineation of what he considers "the basic Marxist insight" on technology is a very Promethean celebration of "the immense productive forces" that must be liberated from their social and political constraints (ibid., 126). But the necessarily revised ontology of technology in Marxist thought, prompted by the acknowledgment of ecologically unequal exchange, requires that we stop sequestering "productive forces" from (global) "relations of production."

An ecological Marxism must exert itself to grasp that the development of the fossil-fueled steam engine is not to be regarded as simply a progressive revelation of politically innocent truths of nature, but the mobilization of various physical and organizational phenomena as components of a complex social strategy ultimately serving to displace work and environmental loads from Britain to African slaves and American soils. Our inclination to distinguish nature from society not merely as analytically

---

[10] This quote from Marx clearly shows how the dichotomy of Nature and Society prevented him from understanding the physical operation of the machine as contingent on unequal exchange.

identifiable *aspects* of material processes – which is incontrovertible – but as two mutually exclusive domains of reality has prompted us to classify technology as fundamentally a product of nature. But technology, we must conclude, is no more "natural" than capitalism.

The Money-Energy-Technology Complex and Ecologically Unequal Exchange

An interdisciplinary theorist struggling to grasp the global ecological impasse of industrial society is faced with a massive and growing literature deliberating on the conditions and processes underlying environmental crisis. Surrounded by a majority of people imbued with mainstream images of the economy and environmental problems, it is understandable that Marxists will want to reconfirm their critical worldview by going back to the foundational nineteenth-century texts of Marx and Engels. However, rather than regressing to the occasionally refractive prisms through which Marx and Engels viewed reality in the mid-nineteenth century, the most significant challenge for an ecological Marxism is to build and expand on their insights, even if this means modifying some components that have appeared essential to a Marxist identity.

An ecologically attuned approach to asymmetric exchange would need to acknowledge that Marx's three volumes on capital are ultimately about the logic of general purpose money, and its insidious capacity to mystify asymmetric exchange. It would need to concede that the progress of the productive forces, rather than something to be celebrated in Promethean fashion, is contingent on an asymmetric exchange of biophysical resources such as embodied energy, materials, land, and labor, and that this asymmetric exchange is made possible precisely by the veil of general purpose money, which makes market exchange seem reciprocal. It would need to see that it is not helpful to phrase the distinction between the material substance of commodities and their market values in terms of use-values that somehow exceed exchange-values, as this inadvertently subscribes to mainstream ideology by suggesting that asymmetric exchange is a matter of insufficient monetary compensation. Finally, it needs to understand that the material, metabolic flows that Marx was struggling to conceptualize are what today would be referred to as embodied biophysical resources, of which labor power is but one example. This means acknowledging that a capitalist surplus does not hinge merely on the appropriation of labor power but can be derived from other economic activities in which income from sales exceeds costs of inputs. For industrial capitalism, this involves purchasing and applying various biophysical resources in the production of commodities yielding economic value. In obscuring asymmetric flows of biophysical resources (including embodied labor), the circulation of money (i.e., exchange-values) makes exploitation possible; this is in complete agreement with Marx's analysis of the exploitation of labor power but proposes to extend the analysis to other kinds of biophysical resources.

We can only speculate on what Marx might have said about such modifications of his theoretical framework, had he been alive today. Possibly, he would have acknowledged that a human artifact such as general purpose money cannot simply be regarded as a reflection of more substantial social processes, but as their very source. He might have observed that technological progress after two centuries of industrialization primarily remains the prerogative of wealthy core areas of the capitalist world-system, and that the accumulation of machinery is the fetishized result of asymmetric global transfers of resources. Hand in hand with this conclusion, he might even have abandoned the conviction that labor power is the only source of surplus value, and enthusiastically embraced the new biophysical metrics – including energy – with which his intuitions about metabolic rifts and asymmetric exchange can be rigorously substantiated. These modifications do not dilute the Marxist critique of capital accumulation, exploitation, and fetishism, but make it more relevant to global concerns in the twenty-first century. The essential dissent vis-à-vis mainstream economics boils down to the issue of whether regular market transactions can be said to involve asymmetric transfers, gauged in terms of the material history or substance of the commodities exchanged, which contribute to economic inequalities and uneven development. A theory that acknowledges the moral and political significance of such asymmetries in international trade represents a crucial counterpoint to the mainstream faith in economic globalization and free trade. Such a theory owes everything to the pioneering insights of Karl Marx, but it is more in line with his spirit to dare to extend the applicability of his revolutionary analysis than to apotheosize his nineteenth-century categories.

10

Subjects versus Objects

Artifacts Have Consequences, Not Agency

There is currently a strong movement in the human sciences to recognize what is often referred to as "distributed" agency. Originally stimulated by Bruno Latour and other proponents of Actor-Network Theory, a fundamental point of this perspective is to reject the "Cartesian" dichotomy between subject and object, in which the human subject is perceived as acting upon passive, nonhuman objects. The alternative, endorsed by Actor-Network Theory and many exponents of the ontological turn in the human sciences, is to perceive the various nonhuman entities with which humans interact as similar sources of agency. This view has proven congenial to several categories of researchers aiming to relativize and challenge traditional paradigms associated with a Western or modern ontology, whether natural science, Eurocentric anthropology, or the anthropocentrism of humanism. Central to what these researchers have in common is the conviction that the Enlightenment view of nature is inextricably tied to colonial European ambitions to dominate the world. Over the past three decades, there has thus been a discursive convergence between STS, postcolonial theory, feminism, and avant-garde ethnography. I refer to the worldview in which these schools of thought converge as *posthumanism*.[1] Their proponents tend to present their perspectives as subversive of the hegemonic worldviews associated with the powerful interests of Euro-American science and technology, siding instead with repressed categories such as indigenous peoples, women, and nonhumans.

---

[1] Kipnis (2014: 44) defines *posthumanism* as "analytic stances that grant agency to nonhuman entities and that downplay the differences between human and nonhuman agency."

This is paradoxical, as the attribution of autonomous agency to inanimate entities such as artifacts has been regarded by Marxists as a capitalist illusion that mystifies relations of unequal exchange. What to Marxists is capitalist ideology to be exposed is thus conceptualized by posthumanists as a subversive claim to be endorsed (Latour 2004a, 2010), and yet both perspectives present themselves as emancipatory. It is not difficult to understand why a social philosopher trained in Catholic theology would want to rehabilitate fetishism (Latour 2010),[2] but more complex to unravel why a generation of social scientists has so eagerly sought to emulate him. There appears to be a widespread confusion between efforts to dissolve hierarchies, on the one hand, and efforts to dissolve analytical distinctions, on the other. In other words, political critique is confused with ontological critique. The dissolution of "Cartesian" distinctions between humans and nonhumans is perceived as a democratic, if not revolutionary, project, but is tantamount to disarming the very possibility of political critique.

Given the recent trend to approach the deliberations on the Anthropocene from posthumanist perspectives (Haraway 2015; Latour 2017), the contradiction between posthumanism and Marxism becomes particularly clear in their divergent understandings of global environmental history. In this chapter, I shall argue that the ontological strategy of the posthumanists leads in a direction diametrically opposite to their professed emancipatory concerns. I shall not engage in exegesis of Descartes, Marx, or even Latour because what they have or have not said is not the issue here, but focus instead on the validity of some central claims frequently encountered in posthumanist ethnography, environmental philosophy, and other strands of social science research. My aim is to question the current urge to dissolve purportedly "Cartesian" distinctions such as between subject and object, society and nature, and human and nonhuman, and to trace its implications for our capacity to analyze and challenge the global power structures with which posthumanists tend to associate them.[3] Precisely because I am in complete agreement with the professed ambition to challenge these power structures, I hope to show that some analytical distinctions that are conventionally dismissed as

---

[2] Paradoxically, theology could provide a solid foundation for critiques of capitalist fetishism, but this is not a direction pursued by the proponents of Actor-Network Theory.

[3] This urge is currently so widespread in the social sciences that I cannot agree with Blaser's (2013: 548) claim that the "assumption of an all-encompassing modernity has come to dominate both scholarly and political analysis to the point that anything that might try to contest it is automatically treated with contempt."

"Cartesian" are in fact indispensable for a truly critical social science. Only by applying such distinctions are we able to grasp the predicament of the Anthropocene and to expose the exploitative global power relations underlying the ideology of economic growth and technological progress.

The anthropogenic global environmental changes that these power relations have generated over the past two centuries have received the attention of several disciplines, such as environmental history and Earth system science, but have not been adequately theorized. I shall begin by considering some of the theoretical options currently offered for understanding the historical relation between human societies and the remainder of the biosphere, arguing that to assemble an integrated theory of global environmental history, we shall have to clarify our position with regard to some contested distinctions commonly referred to as "Cartesian," including our position with regard to posthumanist conceptions of distributed agency.

Options for Theorizing Environmental History

Most of the literature on global environmental change over the past few centuries has been content with describing the observed ecological transformations, without providing any explicit theoretical framework with which to understand them (Hornborg 2010). There has been extensive empirical documentation of processes such as deforestation, biodiversity loss, soil depletion, eutrophication, the spread of invasive species, chemical pollution, changes in land use such as industrialization and urbanization, changes in energy use, greenhouse gas emissions, climate change, and ocean acidification. However, these accounts of biophysical processes have generally not been juxtaposed with social theory. They have either refrained from elaborating explicit social theory or offered rather crude analytical tools such as "economic development," "technological progress," "energy transitions," or "population growth." Such conventional categories reflect mainstream narratives founded on an implicit worldview in which recent planetary processes are understood as consequences of the progression of the human species from less to more advanced conditions. Such "progress" is generally assumed to be beneficial for humanity at large, even when it is conceded that it has troubling ecological repercussions that urgently need to be addressed. To alleviate the various risks generated by human development is viewed as yet another challenge for the advanced economic and technological infrastructures that brought us here – this approach is often referred to as *ecological modernization*. Such understandings of global environmental history do not claim to be grounded in social theory, but rather imply that social theory is not relevant to accounting for environmental change. Whether natural scientists or historians, the researchers who have traced global environmental change have not been equipped with adequate analytical tools for relating such change to societal processes.

Alongside the mainstream trust in human progress, modernization has from its very beginnings provoked more pessimistic narratives envisaging societal decline or even collapse. This set of critical narratives include Marxian interpretations of accumulation as based on exploitation and the related, zero-sum game perspective of world-system analysis (Wallerstein 1974; Chase-Dunn and Hall 1997). In this category I would also include so-called Malthusian concerns with ecological limits to growth and diminishing net energy (Tainter 1988; Tainter and Patzek 2012). Some of the most explicit and elaborate attempts to provide critical perspectives on the trajectory of human societies over the past few centuries, addressing both inequalities and ecological limits, have been offered by ecological Marxists (O'Connor 1994; O'Connor 1998; Foster et al. 2010). I will here be concerned with how the conceptual framework of Marxian social theory relates to the "post-Cartesian" convictions of posthumanism. Before discussing this topic, however, I shall declare my own position regarding distinctions such as subject-object and society-nature and argue for a more restricted understanding of agency than that currently offered by the posthumanists.

Life, Agency, and the Subject-Object Distinction

Let us begin by reconsidering some fundamental categories of the world-view that posthumanists following Latour (1993a) would dismiss as reflecting the obsolete "modern constitution." Although retaining much-maligned "Cartesian" categories such as the distinctions between subject and object, society and nature, and human and nonhuman, the basis for these distinctions, as offered here, is quite different from the dualist ontology of René Descartes. To begin with, I hold that the distinction between living and nonliving entities hinges on the occurrence of *agency*, that is the capacity to act. Agency is propelled by *purpose*. All living organisms have purposes inscribed in their composition, whether the amoeba's absorption of nutrients from its surroundings, the tree's extension of branches into the sunlight and roots into the soil, or a human preparing and ingesting a meal. Such various processes are all examples of agency generated by purposes internal to living beings. When purposes are consciously reflected on, as is often the case among humans, we talk about *intentions*. To attribute agency, purposes, or intentions to nonliving objects is tantamount to fetishism.[4] The purposes that define biotic entities presuppose a certain capacity for sentience and communication (Kohn 2013).[5] Amoebas, trees, and humans are all equipped to register specific aspects of their environments and to somehow respond to them. This capacity for sentience and communication is what defines a *subject*. Abiotic entities such as rocks or artifacts do not have such capacities. They are *objects*. This is an ontological, rather than just a political, distinction.[6]

Anticipating objections from the posthumanist camp, I hasten to provide some qualifications to this contested framework. First, the purposes internalized in living organisms are indeed products of their external relations, whether through phylogenetic or ontogenetic learning. Second, the purposive agency of organisms – most dramatically that of *human* organisms – is indeed partly shaped by and extended through their current engagement with other entities, both living and nonliving. Third, consciousness, reflection, and intentionality are never equivalent to absolute or definitive knowledge, but are a situated, partial, and provisional representation of the conditions of agency.[7] Fourth, subjectivity is a similarly relative concept, encompassing every instance of sentience from the information-processing capacity of amoebas and trees to the emotional life of humans. Finally, we must distinguish between positing the ontological *condition* of subject versus object, on the one hand, and the *perception and/or treatment of entities as* subjects or objects, on the other. To posit the existence of subjects and objects is not tantamount to repressing animals, women, or colonies. Humans and many nonhuman animals can perceive and/or treat an external entity either as a subject with which

---

[4] For a paradigmatic example of such fetishism, see Bennett 2010.

[5] Kohn (2013: 91–92) criticizes Actor-Network Theory and STS for not distinguishing between living, semiotic "selves," on the one hand, and objects and artifacts, on the other, succinctly concluding that "selves, not things, qualify as agents."

[6] To deny the ontological reality of the subject-object distinction may be related to the pervasive psychological denial of the fact that we shall all reach a moment when we are irrevocably transformed from subjects into objects.

[7] The recognition that knowledge is always situated, radically endorsed by posthumanist feminists such as Haraway (1988), is quite compatible with a realist ontology (Bhaskar 1975).

to communicate or as an object with which no communication is possible. This means that both humans and nonhuman animals can make mistakes, treating subjects as objects or objects as subjects. It also means that they can find it pragmatic or instrumental to disregard the sentience of other subjects, as in predation, repression, or some everyday contexts of social life (cf. Buber 1970).

Nonliving objects do not have agency, but they can impact on their surroundings – that is, have consequences for them – in at least four ways. First, they can form physical *constraints* on the agency of living entities, for instance by restricting their movement. Second, they can serve as *catalysts*, prompting them to respond, for instance to weather events. Third, they can be *delegated* specific functions, as in all kinds of human artifacts and technologies. And fourth, they can be *attributed* agency, as when artifacts or other entities are perceived as having autonomous agency that they do not have. This category can be exemplified by objects such as monetary tokens, ancestral mummies, sacred mountains, astronomical bodies, and computers, but also by fetishized living organisms such as sacred trees or divine emperors. In all these cases of "distributed" agency, objects (and living fetishes) may constrain, prompt, or mediate the agency of living organisms. But in no case is it justified to dissolve the crucial difference between purposive agency and merely having consequences.[8] When a fetishized object has the appearance of having agency, it is the perception of the object that influences human agency, not the object that acts.

Symbolism, Semiotics, and the Society-Nature Distinction

The distinction between society and nature, reflected in the division of labor between social and natural sciences, is not merely a modernist conceit serving the interests of European colonialism. The concept of nature has, of course, been endlessly discussed by philosophers, historians, anthropologists, and other social scientists, and no one can claim to have an overview of the vast range of positions taken over the centuries, but because we contest the rejection of the nature-society distinction as a modernist construction, it is appropriate to explain why. Rather than to think of nature as a certain kind of physical things or spaces

---

[8] Although I share Kipnis's (2015: 55) aspiration to distinguish between different kinds of agency, I do not agree that agency should be granted to "anything and everything [that] could affect us."

uncontaminated by humans (cf. Cronon 1995; Soper 1995; Ellen 1996),[9] I regard it as an analytical category encompassing all those aspects of socioecological processes that derive from forces and regularities that *do not require explanations referring to the symbolic capacities of human beings.* "Social," however, refers to those aspects that do require references to symbolism. Nature would thus include, for instance, thermodynamics, gravitation, and photosynthesis, but only the *nonsymbolic* aspects of agriculture, markets, or consumption.[10]

In positing a distinction between society and nature, this view does suggest a variety of what Descola (2013) calls "naturalism," but it disrupts Descola's quadripartite scheme in two ways. First, it posits discontinuities not only with regard to "interiorities" but also among "physical" entities, based on the criterion of whether they are in part generated through symbolic processes. Second, it recognizes a fundamental discontinuity between nonhuman physical entities that are alive versus those that are not. In this view, interiorities and physicalities are as interfused (albeit *analytically* distinguishable) as culture and nature, but nature is discontinuous. Culture or society is distinctly anthropogenic but permeates the physical world, whereas nature encompasses both subjects and objects. While both are justified, the culture-nature distinction does not coincide with the subject-object distinction (Figure 10).

Although increasingly intertwined in empirical reality, society and nature denote aspects that should be kept analytically distinct. This distinction can be expressed as that between *symbolic* and *nonsymbolic* phenomena. Symbolic phenomena such as language are products of uniquely human social processes of negotiating meanings. They are contingent on human subjectivity but are causally influential components in socioecological processes. The social sciences have developed analytical tools and concepts for dealing with human subjectivity (e.g., culture, semiotics), whereas the natural sciences generally have no need for such

---

[9] This essentialized view of nature has been fundamental to the history of wilderness conservation and continues to underlie modern appeals for the preservation of biodiversity, now coupled to the mitigation of climate change (Buck 2015; Wilson 2016; Vettese 2018). The utopian notion of "rewilding" half the Earth not only reproduces the dualistic ontology that places humans outside of nature but also ingenuously suggests, in the spirit of ecomodernism, that intensified production on the "human" half will not encroach on the "natural" half (Büscher et al. 2017).

[10] The symbolic aspects of these phenomena would include, for instance, the cultural food preferences deciding the choice of crops, the socially constructed categories of economics, and the semiotic structures that generate a taste for specific commodities. For a foundational text on the ubiquity of symbolism in human life, see White 1940.

Examples	SUBJECT (Living beings)	OBJECT (Nonliving things)
NATURE (Nonsymbolic phenomena)	Sentience	Rock
SOCIETY (In part symbolically constituted)	Human thought	Artifact

FIGURE 10 The distinctions between subject and object and between society and nature do not coincide

concepts. The social and natural sciences will certainly need to collaborate much more extensively to grasp the hybrid socioecological processes shaping the biosphere, but the traditional division of labor between them reflects undeniable differences between the character of social and natural phenomena.

If symbolic phenomena actively participate in shaping the human environment, it is appropriate to ask to what extent this phenomenon distinguishes humans from other animals (cf. White 1940). The perspective of *ecosemiotics*, going back to the insights of the Estonian zoologist Jakob von Uexküll ([1940] 1982), acknowledges that *all* species engage their environments by representing them and communicating about them in highly specific ways.[11] An ecosystem can in fact be understood as the complex interaction of a great number of subjective perspectives, determined by the sensory and communicative capacities of different species. The physical flows of matter and energy through an ecosystem are only that aspect of ecology that can be registered by natural science; these material flows are *contingent* on the semiotic or communicative flows that are equally diagnostic of ecosystems.[12]

---

[11] This insight is strikingly similar to the perspectivist variety of animism identified among some indigenous peoples of Amazonia (Viveiros de Castro 1998), but it does not necessarily imply that different species literally live in different worlds, as is the conclusion of ontological relativists regarding the radical alterity of nonmodern peoples (Goldman 2009; Blaser 2013; Law 2015).

[12] And vice versa, of course. This is to say that the material and semiotic aspects of ecosystems are *recursively* interdependent, rather than one being ontologically prior to the other.

Type of sign	Relation between the sign and what it refers to
Index	Direct connection (e.g., sound, odor, flavor)
Icon	Similarity (e.g., mimicry, camouflage)
Symbol	Convention (e.g., language, dress)

FIGURE 11 Sign use among humans and other animals

Viewing ecosystems as in part constituted by semiotic flows leads us to ask how different kinds of such flows contribute to ecological change. An ecosemiotic perspective implies that nonlinguistic, sensory signals such as animal sounds, odors, and flavors all participate in the formation of ecosystems, and that this must have been the case throughout the history of life on Earth, billions of years before the emergence of human language. Language added a new dimension to ecosemiosis, most fundamentally by introducing symbolism and culture into the processes by which species transform their environments. Symbols, as defined by Charles Sanders Peirce, are that special kind of signs defined by the *conventional* nature of the relation between the signifier and that to which it refers. The two other major categories of signs are *indices* and *icons*. In the case of the index, the relation between signifier and signified is one of *contiguity* or direct connection, like the odor or sound of an organism or the aroma or flavor of a wild fruit. In the case of icons, this relation is based on *similarity*, like the mimicking of poisonous or unsavory animals known to occur in ecosystems or the camouflage coloring by means of which organisms can avoid detection. The examples I have provided illustrate how such signs – indices and icons – are regular components of ecological systems, generated by processes of natural selection, whether or not there are humans involved (Figure 11).[13] The novelty of human language, from this perspective, is that representation and communication introduce an element of cultural choice or arbitrariness. Rather than being encoded in genes subject to selection, the specifics of linguistic sign systems are the results of cultural processes of innovation and negotiation. Linguistic representations only occasionally mimic the phenomena to which they refer, as in onomatopoeia, but are generally the products of implicit agreement about the meanings of words. The symbolic nature of human

[13] Significantly, natural selection operates on the indexical and iconic signs utilized by nonhuman organisms, but so far not on symbols. In the event that the symbolism with which modern civilization is preoccupied should lead to the extinction of the human species, the latter would no longer be true.

representation is fundamental not only to language but also to cultural phenomena in general. While the general perspective of semiotics provides the basis for a distinction between living and nonliving "actants" (cf. Kohn 2013: 91–92), the specificity of symbolism and culture justify the distinctions between society and nature and between humans and nonhuman animals. The attempts by many posthumanists to dissolve both these distinctions are highly misleading.

To briefly indicate some of the problematic conceptions frequently encountered among posthumanists, we may first note their general inclination to replace analytical distinctions such as those previously outlined with broad homologies or even equivalences that disavow important differences. It is as if the distinctions in themselves are somehow oppressive, provoking deconstruction. To some STS scholars, there is no fundamental difference between the agency of living and nonliving entities, and the distinction between society and nature is presented as a modernist conceit (Latour 1993a). These deconstructions disavow the difference between sentience and nonsentience, and between symbols and other signs. To some feminist scholars, there is no essential difference between humans and nonhuman animals, again disavowing the difference between the symbolic and the presymbolic, and what is disparagingly referred to as "human exceptionalism" (Haraway 2007).[14] In postcolonial ethnography, the inclination is to present indigenous ontologies asserting the animate agency of abiotic entities such as mountains as valid challenges to modern science (de la Cadena 2010, 2015), in effect disavowing the difference between sentience and nonsentience (see Chapter 12). The hyperrelativism of the ontological turn in anthropology and sociology proposes that we should accept radically different ontologies as just as valid as our own, and that their proponents are in fact living in a separate reality in which our Western categories are inapplicable (Viveiros de Castro 1998; Goldman 2009; Blaser 2013; Law 2015).[15] Many have understood the idea of the Anthropocene as evidence of the invalidity of the society-nature distinction (Haraway 2015; Law 2015: 134; Latour 2017). Such attempts by posthumanists to dissolve modern analytical distinctions often appear to derive from genuine aspirations to challenge

---

[14] For a profound defense of human exceptionalism, see Soper (2012).

[15] John Law (2015: 127) has expressed this position with admirable clarity: "Are we simply saying that white people believe one thing, for instance about what we code up as 'nature', whereas Aboriginal people believe something different? Or is something different going on? The new post-colonial response is that the differences are *not* simply matters of belief. They are also a *matter of reals*" (emphases in original).

global power structures but only result in the dissolution of their capacity to do so. Whether its categories in part coincide with those of Descartes, a truly critical social science requires a realist ontology and rigorous analysis.[16]

Marxism and Posthumanism: Fundamental Differences and Possible Dialogue

Actor-Network Theory has frequently been criticized for a disinterest in challenging power and social inequalities (e.g., Bessire and Bond 2014; Martin 2014; Gregory 2014; Kipnis 2015: 54). In part, this appears to be a consequence of its radical empiricism, advocating detailed studies of the interaction of particular humans with particular artifacts (Kipnis 2015). Latour has explicitly rejected macrosociological categories such as capitalism and even society. This has predictably generated a contradiction between Marxism and Actor-Network Theory that seems difficult to reconcile. At the same time, Marxist theory can be criticized for being content with a framework of abstract categories that is rarely anchored in concerns with the operation of concrete social relations and artifacts. The abstract Marxist critique of capital must be concretized in a scrutiny of how the ubiquitous artifacts we know as money and technology contribute to shaping social relations of power and inequality. The radical empiricism of Actor-Network Theory was established through Latour's pioneering studies of the formation of scientific knowledge and technological infrastructure in modern contexts such as laboratories and urban planning (Latour and Woolgar 1979; Latour 1996). The meticulous attention to concrete detail in these case studies is made theoretically interesting through Latour's innovative reflections on the interaction between humans and their various artifacts. Rather than perceiving artifacts as mere extensions of human agency, as was the traditional view, Latour discovered that the artifacts contributed to shaping not only the forms of human agency but also human perceptions. This is the fundamental insight that has guided Actor-Network Theory, a generation of STS scholars, and a great number of researchers in other social sciences such as sociology and anthropology. It provided Latour and his followers with a platform from which to critique traditional understandings of modernity, technology, and scientific knowledge production. However,

---

[16] For sharp illuminations of the contradictions and analytical lapses of Actor-Network Theory, grounded in critical realism, see Elder-Vass (2008, 2015).

although subversive of our trust in science, this critique has not been directed at capital, power, and global inequalities. This latter omission is significant and remarkable, considering that Latour's initial insights on the role of artifacts in human societies emerged in the context of understanding the circumstances of power, dominance, and hierarchy in a comparison of humans and baboons (Strum and Latour 1987). Collaborating with the primatologist Shirley Strum, Latour realized that what had made it possible for humans to extend their fields of social interaction beyond face-to-face relations was precisely their use of artifacts, widely defined to include language and symbols as well as physical implements. Given Marxist concerns, the most obvious human artifact to scrutinize from this perspective is *money*. It seems difficult to explain the disinterest of Latour and Actor-Network Theory in how this uniquely human artifact has shaped, and continues to shape, human social networks extending beyond face-to-face relations and constituting a truly global society.[17] The crucial difference between humans and baboons is capital.

Money is indeed an artifact attributed with spectacular agency. Moreover, it is requisite to the systems of technological artifacts that merge with it in integrating world society. These are the artifacts on which capitalism is founded. It seems inexplicable that a school of thought concerned with how artifacts shape human social relations has neglected to focus on money, and from theorizing its significance for the very feasibility of technological phenomena. By contrast, it is no less remarkable that Marxism has neglected to anchor its theoretical analysis of capitalism in the concrete operation of the fundamental artifacts on which it rests. Capitalism *is* the logic of the everyday operation of money and technology in human societies, but the inherent features of general purpose money generally appear to be as invisible to Marxists as to mainstream economists. Latour says that there is no such thing as capitalism, but why has he not shown an interest in examining how money and technology generate and reproduce human inequalities? Marxists tend to shun the empiricist preoccupation of Actor-Network Theory with individual artifacts, but why do they not systematically scrutinize the very

---

[17] Although Callon (1998) has traced the operation of market logic to the "calculative agencies" employed by market actors, he does not problematize the foundational role of general purpose money in providing a singular, quantitative metric that makes calculation both possible and imperative.

artifact of money?[18] The answers to these questions implicate ideological differences as well as different methodological proclivities regarding modes and levels of abstraction.

Latour would consider Marxists a paradigmatic example of thinkers constrained by the so-called modernist constitution. In their extensive deliberations on the role of nature in capitalist processes, eco-Marxists generally do not lose sight of the analytical distinction between the social and the natural. The particular way in which natural conditions such as thermodynamics are incorporated into Marxist theory is a contested and important analytical issue in its own right, but unlike the posthumanists, Marxists are not inclined to relativize scientific knowledge or challenge the conviction that we all live in a common universe. Sharing such common ground makes it possible to deliberate, for instance, on the relation between the Marxian concept of surplus value and the physical reality of energy. While there are inevitable disagreements on how this relation should be understood, categories of thermodynamics such as energy and entropy are not in question.

In recognizing a distinction between subjects and objects – living and nonliving things – Marxists are also committed to understanding and exposing fetishism. Fetishism is a useful category for the illusion that an object is animate, in Marxian theory exemplified by our inability in capitalism to grasp that our seemingly potent artifacts are ultimately expressions of social relations. Fetishism is the attribution of autonomous agency to inanimate or abiotic things. The externalized interaction of their artifacts – whether money, commodities, or machines – tends to be perceived by humans as determined by the intrinsic properties of the artifacts, rather than by the regulations and features bestowed upon them by human agents. Humans thus become subservient to their artifacts, rather than *vice versa*. Much as players of a board game will refer to the rules, mainstream economists tend to refer to the logic of money and engineers to the logic of their technologies. The extent to which humans are the authors of the social games enacted by their artifacts is obscured from view. The responsibility for human social relations – and for human-environmental relations – is delegated onto things, as if they were the ultimate determinants of society. But the crucial distinction we need to make is between the notion that artifacts such as money have agency and the observation that they simply have consequences. The former is

---

[18] Exceptions include Nelson (1999), Mellor (2005), and Moseley (2016).

fetishism, whereas the latter makes it possible for us to see artifacts for what they are: products of *human* agency that humans could potentially transform. The Anthropocene is ultimately a consequence of the logic of general purpose money, but the design of this artifact – and its consequences – are not inevitable products of human biology.

Anthropocene, Capitalocene, Technocene, Econocene

In an article in *The Anthropocene Review*, Will Steffen and colleagues (Steffen et al. 2015: 91) confirm the argument that Andreas Malm and I made in the same journal a year earlier (Malm and Hornborg 2014), that the designation Anthropocene misleadingly suggests that the global environmental changes it refers to have been propelled by humanity as a whole. What Steffen and colleagues call "the profound scale of global inequality" is clearly reflected in statistics showing that the OECD countries in 2010 accounted for 74 percent of global GDP but only 18 percent of global population. "Until very recently," they observe, the so-called Great Acceleration has been "almost entirely driven by a small fraction of the human population, those in developed countries." However, they continue, this is beginning to change. The most compelling figures they refer to in illustrating this shift are statistics indicating that in 2013 per capita emissions of carbon dioxide in China had surpassed those in Europe. At first glance, such figures do indeed seem compelling, but unless they tell us if the emissions derive from production or consumption, they cannot in themselves be used in support of an assertion that the planetary transformations of the Anthropocene are increasingly being propelled by humanity as a whole. In Sweden, for instance, it has been estimated that total per capita emissions of carbon dioxide, including emissions from the production of all goods *consumed* in Sweden, are about twice those suggested by the official statistics used in comparing the carbon emission performances of different countries. Such figures confirm that current processes of global environmental change continue to reflect unequal relations of power, exchange, and distribution in world society.

This argument is not tantamount to having made the choice to understand the Anthropocene in terms of the history of capitalism *rather than* in those of the history of the human species. For Dipesh Chakrabarty (2014, 2015), seconded by Clive Hamilton (2015), Malm's and my intervention appears to signify such a choice. Hamilton attributes to us an "orthodox Marxian persuasion," objecting that the Soviet Union and Maoist China were no less Promethean than capitalist nations, and that the "broad populace" has willingly collaborated with capital in destroying its own future. Apparently to indicate the inadequacy of Malm's and my position, Hamilton reiterates Chakrabarty's (2009) call for an integration of the history of our species with the history of capital. But this is in fact precisely my point. It is the uniquely human capacity for abstract representation that is requisite to money, and money was requisite to the Industrial Revolution that inaugurated the Anthropocene. To my mind, this account *does* adequately integrate the history of our species with the history of capital. The semiotic capacities of our species made it feasible for us to generate unprecedented intraspecies inequalities. I repeat that no other species could have developed capitalism. To underscore the inequalities underlying recent global environmental change is not to deny that the capacity to develop such inequalities is uniquely human, but Malm's and my point was that to refer simply to the Anthropocene risks leaving the inequalities out of the picture altogether. The establishment and reproduction of a fossil-fueled technology is inherently interfused with social inequalities. The designation of our present time as the Anthropocene might suggest that climate change is the inevitable consequence of how our species is constituted. My objection is that although the potential for capitalism is inherent in our species, it is not an inevitable product of our biology, nor something for which we all have a common responsibility.

It is ironic that the intervention I co-authored with Malm should be interpreted as an expression of orthodox Marxism, considering how concerned I have been to transcend the productivist biases of conventional Marxist theory. To evoke conflicts and material inequalities is not automatically to qualify as an orthodox Marxist. Nineteenth-century steam power was a particularly spectacular example of processes of capital accumulation that had precedents going back several millennia. Alongside the biases of a Promethean trust in technological progress, the most difficult conceptual hurdle for early Marxism was to extricate these concerns with the expenditure of physical energy in production processes from the all-encompassing discourse on monetary profits and economic value. Rather than a genuinely transdisciplinary argument on the relation between energy and money – as presented much later by Nicholas Georgescu-Roegen (1971) – parts of the theoretical edifice of traditional Marxism remain impaired by the aspiration to analytically derive monetary gain from expenditures of energy. This observation was elaborated in Chapters 8 and 9. The C in Marx's succinct formula $M - C - M^{\mathrm{I}}$ signifies a production process that simultaneously increases exchange value and entropy. Although both money and energy are quantifiable – by economists and physicists, respectively – monetary gain is a social construction in the sense that it is contingent on subjective perceptions of human artifacts, while expenditures of energy are physical processes irreducible to human subjectivity. The Anthropocene illustrates how social constructions such as money can have very material consequences, but we cannot hope to grasp or remedy our current predicament unless we retain the analytical distinction between them (that is, between social constructions and their material consequences).

I hope to have shown that, to conceptualize the currently disastrous anthropogenic changes in the biosphere, social theory will need to critically scrutinize the ecological and political consequences of our most celebrated artifacts: money and technology. To do so, we must retain our capacity to analytically distinguish between the symbolic and the biophysical aspects of socioecological processes – society and nature. Only by doing so can we assume the task of redefining economic commensurability. Rather than delegating the destiny of human society and the planet to the insufficiently understood, inherent logic of our entropy-accelerating artifacts, we must achieve societal control over the operation of those artifacts. To be able to regulate global social polarization and environmental change, we must transform the artifacts that generate these processes. Only by simultaneously recognizing the two seemingly incompatible aspects of conventional money – its inexorable material and political trajectories *and* the possibilities of transforming it through democratic processes – can we have a chance of surviving the Anthropocene.

II

Anthropocene Confusions

Dithering While the Planet Burns

In the previous chapter we returned to the observation, in Chapter 2, that posthumanist researchers in the social sciences have been inclined to throw the baby out with the bathwater: their dismay regarding what human society is doing to the rest of the biosphere has led them to reject some fundamental premises of modern science such as the analytical distinctions between the social and the natural, subjective and objective, and semiotic and material aspects of phenomena. In this chapter and the next, I will exemplify how concerns about our deteriorating biosphere can prompt well-intentioned and critical minds to abandon serious analysis of our circumstances. Although I sympathize with their ambition to reject mainstream modern ontology, my objection is that the ontological transformations that we need to undergo should not be tantamount to dismissing science and rigorous analysis. If my tone in this chapter is agitated, it is because I am as dismayed by the ongoing destruction of our planet as Donna Haraway, Anna Tsing, and Bruno Latour, but strongly feel that academics terrified by the predicament of the Anthropocene have a responsibility that goes beyond publishing the hazy and elusive deliberations that are currently referred to as "posthumanism." I am agitated not only because we are destroying the planet but also because legions of critical academics are devoting their intellectual energies to everything *but* contributing to an analytically rigorous grasp of our dilemma. Such a synthesis must necessarily be interdisciplinary. It can only benefit from indignation, but it must not abandon ideals of clarity and analytical rigor.

Like so much else, Donna Haraway has taken her designation for our current historical period, *The Great Dithering*, from a science fiction novel. My dictionary defines *dither* as "a state of nervous agitation or confusion." To the extent that Haraway's book, *Staying with the Trouble: Making Kin in the Chthulucene* (2016), is a reflection of our age, the designation is apposite. Indeed, the Anthropocene gives us plenty of reasons to be agitated, but whether it must lead to confusion is another matter. In reading Haraway's book alongside Anna Lowenhaupt Tsing's *The Mushroom at the End of the World: On the Possibility of Life in Capitalist Ruins* (2015), I am struck by the many things these two influential authors have in common. Both are genuinely indignant about what capitalism has been doing to the world over the past few centuries. Neither of them has any hope that the planet can be saved through some kind of ecomodernist engineering scheme. Tsing frankly realizes that "there might not be a collective happy ending" (ibid., 21). So far we are in agreement. But when it comes to how we best ought to approach the situation, their perspectives raise several serious concerns.

For those of us who keep teaching our students that the very point of writing critical social science is to communicate clear and analytically rigorous arguments, it is frustrating to find them absorbing and trying to emulate the style currently prevalent in posthuman studies. Rather than analytical clarity, the aim of much of this writing seems to be to fashion as imaginative prose as possible, replete with evocative allusions, poetic metaphors, and unbridled associations. The style is personal and anecdotal, the engagement with theory journalistic and superficial. I am not happy about the signals we are sending to our students, who discover that academic success may be inversely proportional to clarity.

But let us consider what is being said. Haraway and Tsing both seem to want to show us examples of social activities that provide some measure of hope, indicating the kinds of projects by means of which we might survive and even transcend the Anthropocene. This is a worthy and respectable undertaking, but the examples they have chosen are dubious, frequently revealing inadequacies in their theoretical approach. In one of her chapters, Haraway provides four examples of "sympoietic worldings" that "nurture well-being on a damaged planet": a collaborative crochet artwork depicting a coral reef, an illustrated book series on the vanishing primates of Madagascar, a video game produced in collaboration with native people in Alaska, and the maintenance and resurgence of sheep breeding and handicraft weaving among the Navajo. What these examples have in common escapes me, as does their alleged capacity to inspire hope. Tsing's example is the ecological encouragement, harvesting, and global trading of a prestigious and expensive mushroom (*matsutake*) for the Japanese market.

In both authors there is a recurrent gap between detailed empirical accounts, on the one hand, and excursions into often obscure theoretical reflection, on the other. Tsing's field experiences among ethnically diverse mushroom pickers in Oregon and landscape conservationists in Japan are elegantly contextualized in terms of the historical dynamics of forest ecology, global migrations of humans and other species, and the shifting economic opportunities afforded by extractive capitalism. She persuasively shows how *matsutake* only temporarily assumes the alienated form of a commodity, when disentangled from the dense social contexts in which its life begins and ends. But the questions pile up. Why is capitalism so often referred to in the past tense, as if we were now experiencing the end of it? Have not opportunities such as the *matsutake* mushroom continuously been appearing (and disappearing) over the centuries? Why should these particular kinds of landscape restoration and harvesting skills give us hope of transcending the Anthropocene?

Efforts to reproduce a specific ecological niche will be proportional to the financial strength of the niche market for which it is requisite. As long as *matsutake* remains a prestigious gift among affluent Japanese, there will be plenty of money to restore pine forests and attract pickers and traders. It remains unclear, however, to what extent this particular commodity chain is relevant to the general destiny of people and ecosystems in South America, Africa, and Southeast Asia. Although diverse in terms of cultural and geographical context, Tsing's studies are almost exclusively conducted within the three wealthiest areas of the world-system: the United States, Japan, and Europe. She thus marvels at the resurgence of forests after deforestation (Tsing 2015: 179) without considering the extent to which the global distribution of such resurgence is contingent on political economy. Even as she comments on the convergence between Japan and Oregon with regard to the abandonment of industrial forestry in both regions (ibid., 205–206), she does not reflect on how this convergence is related to the comparatively high price of land and labor in these economies. Tsing blames the outsourcing of forestry to Southeast Asia on the low price of global timber and the inability of Japan and Oregon to compete, but does not mention that this inability to compete has been determined by the relatively high wages and land rent in wealthier countries. What to Japan and America looks like the failure of the forest industry is really the displacement of environmental loads from these countries to Southeast Asia.

Tsing seems to suggest that human resource extraction is a "good thing" for people and ecosystems. Japanese forest managers convince her that "erosion is good" (ibid., 151) for pine forests and *matsutake*, and she endorses their struggles to remove broadleaf trees and even topsoil to recreate the heavily harvested peasant landscapes that nostalgic Japanese associate with *matsutake* and cherished tradition. Considering the financial significance of the market for *matsutake*, we realize why only *some* "traditional" landscapes are being restored, namely those that are profitable. Reflecting on parallels between China and Japan in terms of how deforested and eroded hills historically gave way to regularly harvested peasant oak-pine forests, Tsing concludes that "those eroded hillsides are the site of a lively regeneration in which oak, pine, and matsutake have a good thing going – not just for peasants but also for many kinds of life" (ibid., 189). Now that the peasants are gone, we must recall, the only thing that preserves such landscapes is money. But the profitability of a prestigious mushroom is no apology for deforestation, and very far from a representative example of livelihoods opened by environmental degradation. Although Tsing cites an environmental economist in Kyoto who asserts that "future sustainability is best modeled with the help of nostalgia" (ibid., 182), nostalgia alone clearly does not shape our landscapes, or the Global South would be a very different place.

Haraway frequently refers to Tsing, and Tsing is obviously inspired by Haraway. Both draw conclusions from their case studies that seek to fuse an ecological critique of capitalism with excursions into specialized branches of biology. The average anthropological reader will be asking for a more profound theoretical engagement with capitalism, on the one hand, and evidence of the robustness of their biological deliberations, on the other. To rethink ecology through the lens of a nonmodern ontology is a supremely valid anthropological pursuit (e.g., Kohn 2013), but when an anthropologist personally deliberates on contested frontiers of natural science, it is less persuasive. Haraway (2016: 65–66) does not hesitate, for instance, to discuss molecular and comparative genomic research on choanoflagellates (*Salpingoeca rosetta*) or the implications of pea aphid symbiosis with *Buchnera*. When Tsing (2015: 236) cites research finding that "single nucleotide polymorphisms (SNP) are good for population-level differentiations," we similarly wonder if this is a topic to be discussed by anthropologists. When she speculates about the possibility that mushroom spores "blow around the earth" in the stratosphere, it seems to be mostly just to provide her with "the pleasure of thinking: the spore-filled airy stratosphere of the mind" (ibid., 228). How "happy it feels to fly with spores and to experience cosmopolitan excess," she exclaims (ibid., 238).

Like Haraway, Tsing has an urge to rephrase familiar thoughts in a more poetic, if less accessible, jargon. Foraging "might be considered dance" (ibid., 242). Traditional agroecology is the "polyphonic assemblage" of the "rhythms" of different crops resulting from "world-making projects, human and not human" (ibid., 24). To collaborate and communicate, within or across species, is to be "contaminated by our encounters" (ibid., 27). Consequently,

> contaminated diversity is everywhere.... If a rush of troubled stories is the best way to tell about contaminated diversity, then it's time to make that rush part of our knowledge practices.... It is in listening to that cacophony of troubled stories that we might encounter our best hopes for precarious survival. (ibid., 33–34)

Striking a good bargain on the market, in Tsing's parlance, is an act of "translation" (ibid., 62). Capitalism "has the characteristics of an assemblage," in which commodification and alienation are forms of "disentanglement" (ibid., 133). Assemblages, in turn, are "performances of livability" (ibid., 157–158). Although aware of the dangers, she finds it insightful that biological symbiosis is commonly referred to as "outsourcing" (ibid., 143–144). In this neoliberal idiom – which represents capitalism as "performances of livability" – it is not surprising to find ecological disturbance assessed as "ordinary" (ibid., 160) and precarious living as "always an adventure" (ibid., 163). Although precarity means "not being able to plan, ... it also stimulates noticing, as one works with what is available" (ibid., 278). Tsing is aware that the freedom-loving, culturally "disentangled" entrepreneurs who are willing to live with precarity – without wages, benefits, or universal standards of welfare – are ideal participants in globalized supply chains (ibid., 106), but she does not seem to see the neoliberal reinterpretation of capitalism as sinister in itself. The fragmentation of protest against capitalism dissolves the "urge to argue *together*, across the viewpoints emerging from varied patches, about the outrages of accumulation and power," yet "this is not the end of politics" (ibid., 134).

> "Without stories of progress, the world has become a terrifying place. The ruin glares at us with the horror of its abandonment," Tsing concludes. "It's not easy to know how to make a life, much less avert planetary destruction," but we "can still explore the overgrown verges of our blasted landscapes." (ibid., 282)

Tsing's somewhat impalpable recipe is "a politics with the strength of diverse and shifting coalitions – and not just for humans" (ibid., 135).

Although its metaphors and reflections are frequently hazy, Tsing's book pursues a consistent argument about the unpredictable conditions of economic survival in a globalized capitalism responding to the historical vicissitudes of cultural niche markets and biologically complex landscape transformations. It is difficult to find a similarly coherent theme in the way Haraway (2016) frames the sprawling fancies on which her diverse chapters are based. The designation "Chthulucene"[1] is as far-fetched as it is difficult to pronounce. It is introduced to "name a kind of timeplace for learning to stay with the trouble of living and dying in response-ability on a damaged earth" (ibid., 2). Phrases such as these have not helped me to understand what point Haraway is trying to make, yet they recur throughout the book. Given that she repeatedly emphasizes the centrality of collaboration across all kinds of boundaries (often marked by adding "-with" to a verb), it is difficult to understand why she has not exerted herself more to communicate her ideas to readers like me. Her exertions seem instead to aim at maximum unintelligibility and inaccessibility. This is not to invite collaboration. When she introduces the book by explaining that her "fabulated multiple integral equation for Terrapolis is at once a story, a speculative fabulation, and a string figure for multispecies worlding" (ibid., 10), I am immediately lost. And thus it continues. The system of idiosyncratic tropes that organizes her text is no doubt "original" (as the blurb says), creative, and perhaps poetic, but clearly not intended to communicate an argument. Page after page, Haraway embarks on sentences that do not convey decipherable messages, but that delegate sense making to the play of free and frequently unfathomable associations. I find myself paying more attention to the fact that identical sentences are repeated here and there than to their significance. After carefully reading her book from cover to cover, I am unable to think of any coherent argument that it has made. To the extent that its aspirations are poetical rather than analytical, a reader must ask if it is at all feasible to summarize a poem. How do you argue with a poet? The terror of the Anthropocene can obviously inspire poetry as well as analysis, but poems alone will not suffice to guide students who hope to engage in research or political activism. As I am well aware of how tremendously influential Haraway's contributions to anthropology tend to be, my expectations of her approach to the Anthropocene were as great as my disappointment.

Haraway (ibid., 5) acknowledges her "partners in science studies, anthropology, and storytelling – Isabelle Stengers, Bruno Latour, Thom van Dooren, Anna Tsing, Marilyn Strathern, Hannah Arendt, Ursula Le

---

[1] The word *Chthulucene* is in part inspired by a monster in a 1926 fantasy story by H. P. Lovecraft evoking the indifference of the universe to the fate of humankind.

Guin, and others." Indeed, references to Stengers, Latour, Tsing, and Strathern recur throughout her book, and their blurbs decorate the covers (as Latour, Strathern, and Le Guin have also provided blurbs for Tsing). This inner circle of "tentacular" thinkers (ibid.) do seem to appreciate each other's styles of thinking and writing (but I doubt that Arendt would have reciprocated the acknowledgment). Their current prominence within some strands of anthropology makes me wonder if the Anthropocene demands nothing less than a cognitive rewiring of the human species – the replacement of ideals of clarity and seriousness with a relaxed and sensuous flamboyance – or if that prominence is itself symptomatic of the imminent collapse of our global civilization. Perhaps both are true. While I would be the first to agree that the restricted forms of rationality of economists and engineers are responsible for the Anthropocene predicament, I continue to hold that it is more promising to respond to the threat of apocalypse by rigorously challenging those forms of rationality than by abandoning serious analysis altogether. I do not believe that the "tentacular" thinking that currently is so popular in posthuman anthropology could possibly qualify as intellectual progress.

Haraway's chapters sprawl in diverse directions, apparently pursuing haphazard fancies rather than a coherent theme. A chapter called "Awash in Urine" explores various far-flung implications of the estrogen treatment that her aging dog receives for leaking urine in her owners' bed. The book finally turns into pure science fiction, as the author indulges her personal fantasies about world developments four centuries into the future. Haraway's scenario emphasizes voluntarily reduced human reproduction, leaving a planetary population of only three billion in 2425. More remarkably, a growing proportion of this population are genetically engineered, cross-species "symbionts" combining human and nonhuman DNA. The central personality in the story, a sequence of five linked and overlapping lives all called "Camille," is thus a hybrid between a human and a monarch butterfly. The world is organized into "Communities of Compost," and Camille is born into such a community named New Gauley, West Virginia. The year is 2025, midway through the period that future historians will call the Great Dithering. In such communities, Haraway (2016: 150) imagines,

Decolonial multispecies studies (including diverse and multimodal human and nonhuman languages) and an indefinitely expandable transknowledging approach called EcoEvoDevoHistoEthnoTechnoPsycho (Ecological Evolutionary Developmental Historical Ethnographic Technological Psychological studies) were essential layered and knotted inquiries for compostists.

It struck me that Haraway does not devote a single word to New Gauley's economy or subsistence base. How does this imagined future community combine super-advanced biotechnology with a lighter pressure on the planet's resources? How does it feed its inhabitants? Although Haraway and Tsing share a passionate interest in nonhumans – Haraway calls herself a "lifelong animal lover" – it is evident that their practical engagement with other species does not extend much beyond owning dogs and picking mushrooms. For those of us who have spent decades in the countryside raising animals, growing crops, and caring for a forest, the notion of multispecies entanglement is more than intriguing jargon. Although it may sound innovative to academics comfortably at home in their urban offices, for many rural people it has always been obvious that sheep, mice, trees, and weeds have purposes and agency. I am reminded of the ontological gap I sensed already as a graduate student between my urban colleagues in anthropology and my rural neighbors dedicated to farming and forestry. The anthropologists' reflections on the exotic life-worlds they had encountered during fieldwork on distant continents often sounded a great deal like rural experiences anywhere. Their cosmopolitan, middle-class gaze frequently focused precisely on the embeddedness – among familiar persons and landscapes – of which urban modernity had deprived them. I have often been struck by the same reflection: much of what ethnographers seek, discover, and articulate in remote places is more about crossing social distances than geographical or cultural ones. Reading Tsing and Haraway, I have frequently wondered to what extent their neologisms and sense of discovery derive from an urge to verbalize a mode of experiencing the world that lies just beyond their reach but remains fundamental to local people all over the planet.

To the extent that the abandonment of serious analysis would be a symptom of the collapse of our exploitative global civilization, I would not have any objections, as this might halt the Sixth Extinction (Kolbert 2014) and other ongoing processes of planetary destruction. Unfortunately, however, the growth enthusiasts and ecomodernists who are promoting this civilization are unlikely to be the least perturbed by posthumanism. In keeping critical human science defused – preoccupied with crochet artwork, leaking dogs, and expensive mushrooms – the promotion of posthumanist discourse is ultimately tantamount to looking away while neoliberal capitalism continues to destroy the planet. In other words, it can only serve as a convenient accomplice of neoliberalism.

Haraway and Tsing both express critique and distrust of a loosely defined phenomenon of capitalism. Haraway repeatedly refers to geographer Jason Moore's efforts to champion the concept of the "Capitalocene," and she generously endorses his monograph *Capitalism in the Web of Life: Ecology and the Accumulation of Capital* (Moore 2015). Are we witnessing the unlikely convergence of posthumanism and Marxism? A major chapter from Haraway's *Staying with the Trouble* is included in Moore's (2016a) edited collection *Anthropocene or Capitalocene? Nature, History, and the Crisis of Capitalism*, and Part I of Moore's collection is even subtitled "Toward Chthulucene?" Moore's original aim appears to coincide with that of the conference World-System History and Global Environmental Change, which I convened in Lund in 2003, and to which he was invited, that is, to bring together insights on the world-system and the Earth system (Hornborg and Crumley 2007; Hornborg, McNeill, and Martinez-Alier 2007). Both Haraway and Moore acknowledge that the word *Capitalocene* was invented by my former graduate student Andreas Malm at a seminar in Lund in 2009, while Moore was employed as lecturer at our Human Ecology Division. However, the analytical implications of this concept for Moore appear to have been significantly different from those drawn by Malm and me. Although we are all critical of the way mainstream discourse on the Anthropocene projects an image of the human species (rather than a privileged global minority) having transformed planetary biogeochemistry, which to Malm and me means couching a primarily societal predicament in the idiom of natural science (Malm and Hornborg 2014), Moore argues for a dissolution of the analytical boundary between the social and the natural. This concession to the posthumanist fads championed by Haraway's and Latour's cohort of "tentacular" thinkers is not only completely at odds with historical materialism and Marxist theory – as evident in a rudimentary reading of Latour – but also, in my view, dismantles any chance of politically challenging the destructive forces ravaging our planet. Moore's deliberations signify a posthumanist co-optation of the critique of capitalism, which serves no other interests than those of neoliberalism.

The diverse styles and outlooks of the different contributors to Moore's (2016a) collection do not add up to a coherent approach to current global dilemmas. They range from the clear and cogent chapters by Eileen Crist and Elmar Altvater to the hazy and sometimes unintelligible reasoning of Haraway and Moore. Crist's (2016) chapter "On the Poverty of Our Nomenclature," previously published in the journal *Environmental Humanities*, powerfully exposes the insidious vocabulary of Anthropocene discourse, "so matter-of-factly portraying itself as impartial and thereby erasing its own normative tracks even as it speaks" (ibid., 18). She points to the glaring contradiction between the hubris of our all-powerful species naming an epoch after itself and aspiring to control the planet, on the one hand, and our catastrophic incapacity to control our demography, economy, or technology, on the other. To Crist, the contemporary rhetoric on the integration of the social and the natural conceals the "assimilation of the natural *by* the social" (ibid., 28; emphasis in original). In retaining the distinction, this interpretation provides a welcome antidote to Moore's and Haraway's blurring of the analytical boundaries. Altvater's (2016) chapter "The Capitalocene, or, Geoengineering against Capitalism's Planetary Boundaries" similarly deplores how nature in modernity is "torn from its natural context and integrated into an economic circuit of value circulation" (ibid., 149). This is not equivalent to saying, in a literal sense, that "the external world – what Marx calls external nature – is a creation of capitalist modernity" (ibid.), but that the societal subsumption of nature is disguised by the *representation* of nature as detached from society. Altvater correctly observes that the technological augmentation of labor productivity – and of the *relative* production of surplus value – during the Industrial Revolution, which to Marx and Engels signified "a rupture in human history," crucially derived from the harnessing of inorganic energy (ibid., 143–146). This point should hardly be controversial (cf. Altvater 2007; Huber 2008; Malm 2016), yet, curiously, the significance of fossil energy is questioned in Moore's own chapter.

Moore's (2016b) chapter "The Rise of Cheap Nature" summarizes his outlook in *Capitalism in the Web of Life*. To redefine *capitalism* as "neither a purely economic nor social system," Moore (ibid., 81) resorts to Haraway's assertion that it should be viewed as "a historically situated complex of metabolisms and assemblages." This is neither a contradiction nor a clarification vis-à-vis Marx. Like Haraway, Moore muddles the indispensable analytical distinction between the social and the natural, without which any critique of capitalism would be impossible. He then goes on to challenge the "fossil capital narrative" of the Industrial Revolution, tracing capitalism's "fundamentally new law of environment-making" (ibid., 89) to the fifteenth century. Moore asserts that

the rise of capitalism after 1450 marked a turning point in the history of humanity's relation with the rest of nature. It was greater than any watershed since the rise of agriculture and the first cities. And in relational terms, it was even *greater than the rise of the steam engine.* (ibid., 96, emphasis in original)

This is not a persuasive way of writing environmental history. In the sense of material infrastructures recursively reproduced through symbolically constituted exchange relations, preindustrial forms of capital accumulation have occurred for millennia; the most significant historical discontinuity was indeed the eighteenth-century harnessing of fossil fuels as a source of mechanical energy. As Altvater (2007, 2016) and others have made clear, it was the unprecedented increase in labor productivity afforded by fossil-fuel technology that created modern capitalism and the predicament of the Anthropocene (cf. Malm 2016). The destructive environmental history of preindustrial European expansion traced by Moore is well known, but the global requisites and repercussions of the Industrial Revolution require a theoretical framework that transcends his account of the kind of social organization established in Europe in the fifteenth century. In its Eurocentrism, this aspect of his argument is very difficult to reconcile with his aspiration to provide a global account of the emergence of capitalism.

Moore's most fundamental muddle pertains to the appropriation of what he calls the "unpaid work/energy of global natures," which highlights the "unity of human and extra-human work" hitherto enveloped in a "Cartesian fog" (ibid., 89). Certainly, as I have argued for decades – ironically contradicted 19 years ago by Moore (2000) – labor power as conceptualized in classical Marxism is not the only productive resource that is asymmetrically appropriated, and energy is indeed a common denominator of these appropriated resources. Furthermore, such appropriation is an essential requisite of technological progress, which is nevertheless understood in Marxist thought merely as an increase in labor productivity contingent on the inexorable (and politically innocent) advance of the productive forces. But Moore's emphasis on the conversion of nature into advancing labor productivity does not lead him to critically theorize the phenomenon of modern technology as a global social strategy of appropriating nature to physically establish social inequalities. As for so many others concerned with the ecological dimensions of capitalism, Moore's dilemma is how to reconcile global environmental inequalities with classical Marxist theory, particularly its labor theory of value and its technological optimism. The labor theory of value is misguided in aspiring to analytically derive exchange-value from inputs of labor energy, technological fetishism in not recognizing that technological progress is contingent on asymmetric transfers of resources. Although at times sounding like he wants to replace the Marxian labor theory of value with an equally misguided energy theory of value, Moore chooses to evade the conundrum altogether by escaping into the conceptual haze of posthumanism. "Capitalism's metabolism of work/energy is crucial," he writes, "because it sharpens [*sic*] our focus on how human work unfolds through the *oikeios*: the pulsing, renewing, and sometimes-exhaustible relation of planetary life" (ibid., 90). While such phrases no doubt appeal to the likes of Haraway, I cannot see how they sharpen our focus.

Moore's argument that much of the work/energy appropriated by capital is "unpaid" is difficult to distinguish from the assertions of mainstream economists about environmental externalities, undervalued "true costs," and unpaid ecosystem services. Conventional Marxism thus continues to be conceptually constrained by the monetary bias of the economic system that it challenges: as I argued in Chapter 9, to use concepts such as "value" and "cost" is to implicitly assume a society organized in terms of general purpose money. Altvater (2016: 148), like Christian Parenti (2016: 167–169) in his contribution, joins Paul Burkett (1999) in understanding Marx's concept of "use-values" as referring to biophysical qualities that should be distinguished from the monetary exchange-values that conceal them, but to conceptualize biophysical nature in terms of value is to confuse physics and economics. Such tortuous struggles of heterodox economists to integrate the social and the natural are as futile as the attempts of neoclassical economists to deal with environmental degradation. It requires a great deal of analytical effort to expose the historical and continuing appropriation of embodied labor and ecological resources by wealthier parts of the world-system, but connections are increasingly being made between the theory of ecologically unequal exchange and the dilemmas of the Anthropocene (e.g., Roberts and Parks 2008; Jorgenson and Clark 2009; Malm and Hornborg 2014; Bonneuil and Fressoz 2015). By now it has become widely recognized that the disastrous ecological trajectory of global society is inextricably connected to its widening inequalities. The missing link in fully grasping what I call the money-energy-technology complex, however, is a necessary reconceptualization of modern technology as a global strategy of physically redistributing work and environmental loads. It is the very ontology of technology that is at stake. Rather than merely a category of magical ingenuity, technology is the link between our planetary overshoot and the increasing polarization of rich and poor.

Moore's intuitions about a capitalist "world-ecology" do not provide an analytically rigorous account of the money-energy-technology complex. He does not explain how the artifact of money is the root of asymmetric exchange and requisite to modern technology, nor does he allow his notion of "unpaid work/energy" to explicitly confront the labor theory of value. In not grasping how nineteenth-century Marxian value theory is ultimately founded on the monetary framework of the society it aspires to challenge, he is unable to provide an analytically coherent account of the relation between the social and the natural. Money-based concepts of value that are derived from capitalist society and ideology must be kept analytically distinct from labor energy and other biophysical resources of nature to understand how society and nature, through asymmetric exchange, are interfused in technology. To simply dispel the distinction between society and nature is not at all helpful in our struggles to grasp the current global predicament.

The currently widespread urge to abandon an analytical distinction between the social and the natural can often be traced to Bruno Latour's (1993a) seminal book *We Have Never Been Modern*, which argues that the society-nature distinction is an ideological prop for modernity. In *Capitalism in the Web of Life*, Moore indeed refers to Latour, apparently unaware that the latter has explicitly denied the very existence of capitalism. As we saw in Chapter 2, Latour's prolific deliberations on the Anthropocene tend to dwell on its alleged monist implications, supposedly contradicting modernist ideology by decisively dissolving the distinction between society and nature. In his view, apparently, this is equivalent to verifying the acknowledgment of agency in nonbiotic things such as geological formations and greenhouse gases. In Latour's (2014) Holberg Prize Lecture, "Agency at the Time of the Anthropocene," given in response to what he refers to on his website as the equivalent of a Nobel Prize in the humanities, he argues that the agency of general Kutuzov in Tolstoy's *War and Peace* is comparable, if not identical, to the agency of the Mississippi River when it breached its banks in 2005. This argument has several absurd implications for moral philosophy, law, the humanities, and human social life in general. It implies, for example, that a hermeneutic perspective is completely irrelevant and that rivers are as accountable for their "actions" as army generals. The most problematic implication of Latour's ambition to dissolve the subject-object distinction, however, is finally not the fetishistic attribution of agency to nonliving entities, but the withdrawal of responsibility and accountability from human subjects. The denial of accountability in human subjects – accomplished by putting them on a par with nonhumans – is quite congruent with the relinquishment of responsibility that is implicit in the posthumanist stance of Latour and his followers. The uniqueness of human responsibility – which simply cannot be extended to rivers, volcanoes, or even dogs – remains an insurmountable dilemma for posthumanism. When Haraway (2016: 29) asserts that "[w]e are all responsible to and for shaping conditions for multispecies flourishing," the humanist must ask who or what she includes in the category of "we."

As clarified by the environmental historians Christophe Bonneuil and Jean-Baptiste Fressoz (2015: 75), Latour and others have subscribed to "the overly simple thesis according to which modernity has established a great separation between nature and society, a separation that allegedly prevented us from becoming aware of ecological issues, and that was only challenged quite recently." The claim that the interfusion of nature and society has been ignored in modernity is historically false (Locher and Fressoz 2012). Given the historical evidence of "a very acute awareness of the interactions between nature and society" throughout the Industrial Revolution and the entire modern period, Bonneuil and Fressoz (2015: 76–82) observe that such a misleading view tends to depoliticize environmental issues in the past. Latour's crusade against dualism is not only historiographically flawed but also misleadingly tends to conflate the distinction between society and nature with the distinction between subject and object, and the latter with the distinction between a calamity and its victims. He thus attributes to the Anthropocene an "utter confusion between objects and subjects" (Latour 2014: 9). But the distinction between the social and the natural is not synonymous with the distinction between the subjective and the objective. As I argued in the previous chapter, there are *social objects* (artifacts) and *natural subjects* (nonhuman organisms), of which only the latter have purposes and agency. Latour insists that even planets and rivers have "goals" just as humans and other living organisms do, but he is quite mistaken to interpret global warming as a

complete reversal of Western philosophy's most cherished trope, [through which] human societies have resigned themselves to playing the role of the dumb object, while nature has unexpectedly taken on that of the active subject! (Latour 2014: 13)

Global warming is no more nature's purposive revenge on human society than was the Dust Bowl of the 1930s or any other environmental disaster over the course of human history. Nor should the relation between calamities and their victims be confused with the ontological distinction between sentient subjects and nonsentient objects. To distinguish between sentience and nonsentience is not to comment on – let alone justify – any infliction of harm, whether by humans on nonhumans, humans on other humans, or nonhumans on humans.

Such analytical confusion makes me puzzled by Latour's professed influence on a vast number of colleagues in the human sciences, many of whom attempt to emulate his style but would have serious problems persuasively reiterating his arguments. As I concluded regarding Haraway, Latour's primary aim is obviously not to communicate clear lines of reasoning. His mission is not to help his readers grasp what he is saying. A professional expert on the building of alliances to promote particular discourses, Latour has excelled in putting such insights into practice. Most importantly, I am disturbed by the absence in his work of political positioning. I have vainly searched his texts for an indication of some observation that could be regarded as subversive of the neoliberal world market that continues to generate the obscene inequalities, environmental degradation, and financial instability of the Anthropocene. Even Latour's (2004a, 2004b) most explicit attempts to elucidate how his approach might be conducive to criticism are characteristically obscure and evasive. Although fans like Graham Harman (2007) find it strange that Latour's "impact among philosophers has so far been minimal," it may be because philosophers tend to adhere to incontrovertible criteria for analytical rigor.

Latour's creative but frequently unintelligible streams of thought have contributed some gems of insight, but not a coherent perspective on the Anthropocene. He has observed that artifacts are what distinguish human societies from those of baboons and that the specific features of our artifacts are significant for the trajectories of our societies, but he has ignored the "agency" of the artifact of money. He has clarified how misguided it is to "purify" hybrid, socionatural phenomena as *either* social or natural, which is indeed supremely true of the sequestration of the modern categories of economy and technology, but is mistaken in wanting to jettison the analytical distinction between nature and society altogether. Mainstream economists indeed appear to believe that their accounts of economic progress have no need for nature, while mainstream engineers appear to believe that their accounts of technological progress have no need for world society. Yet nature and society are not figments of modernist imagination. The modernist trust in an economy sequestered from nature and a technology purified from society is what has brought us the Anthropocene.

12

Animism, Relationism, and the Ontological Turn

In the previous two chapters I have critically discussed so-called posthumanist approaches in social science, and I have mentioned that much contemporary anthropology is committed to an outlook frequently referred to as the ontological turn. In a nutshell, most proponents of this outlook would argue that different groups of people may literally live in different worlds. In their view, there is no single natural reality to which our various perspectives refer, but as many realities as there are perspectives. They would argue that to posit a scientific approach to reality as more valid and reliable than traditional, nonmodern ontologies such as animism is tantamount to adopting a condescending, top-down relation affiliated with European colonialism. While such condescension and outright contempt for non-European worldviews have been abundantly illustrated in the history of scientific thought, the indiscriminate rejection of Enlightenment concepts of reason and truth cannot be an appropriate response. However, my approach to the far-reaching social consequences of artifacts such as money and technology suggests a concession that there are aspects of an object-oriented posthumanist ontology that deserve to be integrated into social theory. It is because of this conundrum that I devote so much space to the deliberations of posthumanists on the role of nonhuman objects in social life. As we have seen, there are indeed fundamental aspects of a "modern" outlook that implicate colonialism and should be rejected, but paradoxically, the identification of such points of symbiosis between Enlightenment science and colonialism is best accomplished precisely by applying Enlightenment concepts of reason and truth. While I share the anticolonial aspirations of so-called political ontology (Blaser 2013), I do not think that radical ontological relativism will help to expose the power structures and asymmetric resource flows that continue to underpin neocolonialism today.

To substantiate this point, I shall first probe and discuss a prominent example of anthropology conducted with this relativist outlook as point of departure – Marisol de la Cadena's (2015) account of her struggles to understand the radically unfamiliar world inhabited by a family of Quechua-speaking agro-pastoralists in Peru – and then turn to the anthropology of Tim Ingold, whose approach also challenges the ontological outlook of mainstream modern people, but whose ambition to explore actual features of a single world shared by all humans is not as conducive to relativism. Ingold does not share the conviction that humans can literally inhabit different worlds, but he suggests that modern people tend to perceive their environment in a distorted way and that the perceptions of nonmodern people can give us clues about a more valid way of understanding human-environmental relations.

In discussing their quite different challenges to modernity, I will pay particular attention to three aspects of their perspectives: how they understand animism, how they represent human-environmental relationism, and whether their approaches can be used to confront the power structures that threaten the environment. What de la Cadena and the post-humanists discussed in the previous chapter all have in common is a critique of modernity that is founded on a championing of nonmodern perceptions and convictions. Latour, Haraway, Tsing, and de la Cadena are hugely influential in anthropology and belong to the core of the field's "ontological turn." Although I share their critique of modernity, I do not agree that an effective challenge to it can be founded on the worldviews that continue to succumb to it. Having worked with indigenous people in several parts of the world, I am as dismayed as these anthropologists are by the fact that nonmodern people tend to be unable to halt the relentless expansion of modernity.[1] This observation is definitely not a matter of condescension. To deny this inability to resist modernity does not help us to understand it. Instead of championing the kind of nonmodern ontologies subsumed under the concept of animism, my argument on the logic of the money-energy-technology complex applies analytical but transdisciplinary thought. It shares the general anthropological readiness to

---

[1] It is not a coincidence that the mainstream modern image of authentically "indigenous" people tend to assume the criterion that these people live in societies that are not dominated by general purpose money.

deconstruct our modern conceptions as cultural constructions but retains the aim to reconstruct a convincing narrative building on rigorous analysis.

Convictions, Beliefs, and the Suspension of Disbelief

What kinds of perspectives on human-environmental relations are championed by proponents of a so-called object-oriented ontology? Marisol de la Cadena (2015) has made an earnest and sympathetic attempt to transcend the various boundaries that separate her own biography and outlook from those of her ethnographic interlocutors. She shares with her readers the various reflections raised by her struggles as an urban intellectual to comprehend the lifeworlds of Quechua-speaking agro-pastoralists (*runakuna*) inhabiting the rugged rural landscapes east of Cuzco. The project of adequately understanding their experience of the world proves indistinguishable from the challenge of formulating an anthropological position that does not simply adopt a top-down documentation of their "beliefs," but grants their reality an ontological validity that serves as a critique of mainstream modern worldviews and their interfusion with power.

Ultimately, de la Cadena's deliberations concern the role of anthropology in the ongoing shift from colonial to neoliberal perspectives on cultural diversity. She highlights the continuities linking sixteenth-century Spanish extirpation of Andean "idolatry" to a modern Peruvian president's dismissal of animism as an "absurd" and "primitive" form of religion posing an obstacle to development. But the assumption of enlightened superiority, she observes, is as evident among the very leftists who would criticize the president's intolerance of religious and cultural diversity. The long-standing leftist struggles in Peru have consistently been conducted within the confines of Enlightenment discourse, even when the land reforms they endorsed were largely propelled by rural indigenous populations subscribing to a very different ontology. Mariano Turpo exemplifies such political agency, allied with the urban left but experiencing the struggle in distinctly different terms. The close attention to Mariano and his son Nazario Turpo's understandings of their interactions with the outside, non-*ayllu* world is no doubt de la Cadena's most significant achievement.

If the colonial and modernist repression of diversity represents a blunt and straightforward form of power, we also read between de la Cadena's lines about the subtle and insidious power of neoliberalism: the official celebration of cultural diversity through tourism, museum exhibits, and the recognition of indigenous rights. But if the logical historical trajectory of colonialism was a revolutionary leftist struggle for emancipation, the political trajectory of neoliberalism remains obscure. De la Cadena (2010) and Blaser (2013) suggest "cosmopolitics" and "political ontology," but to the extent that such political struggles require a posthumanist foundation, their proposals raise a number of objections.

The dilemma for an emancipatory anthropology is no longer how to confront colonial or modernist aspirations for cultural uniformity, but how to confront neoliberalism. Although it may seem democratic to abandon a realist ontology and officially accept one's interlocutors' claims as valid, no matter how counterintuitive, such public suspension of disbelief is hardly conducive to one's own political positioning. To propose that the anthropologist's hosts literally live in a different world than their guest (cf. Law 2015) may be a convenient way of expressing respect – while circumambulating differences in class and purchasing power – but is tantamount to an extreme form of exoticism that disqualifies the common parameters through which power can be gauged and challenged. Nor, it seems, was political ontology even an efficient strategy for rescuing the mountain Ausangate from being transformed into a pile of rubble. As de la Cadena (2015: 275) observes, "[T]o save the mountain from being swallowed up by the mining corporation, activists themselves – *runakuna* included – withdrew *tirakuna* [earth-beings] from the negotiation."

Remarkably, however, de la Cadena asks her readers to take seriously not only the relationships that *runakuna* maintain with earth-beings but also the very existence of those earth-beings. The former is a crucial and obvious foundation for the ethnographic project since its inception, while the latter has become a shibboleth for the ontological turn in anthropology. If we are seriously prepared to endorse animistic mountain worship, to the point of deploring the exclusion of earth-beings from the public policy discourse that saved Ausangate from a mining project, what are the political implications for anthropology? What is the significance of animism for our endorsement of a land reform or environmental protection? If the discursive eclipse of "earth-beings" by "defense of the environment" is a deplorable accommodation to modernity, what alternative discourses would have been preferable from the perspective of political ontology? To the many anthropologists for whom a geological formation must remain devoid of sentience and purpose, local assertions about the agency of Ausangate, while certainly deserving acknowledgment, respect and sensitive translation, cannot be taken literally.

Translation necessarily implicates the anthropologist's own personal biography and frame of reference. Urban academics are predictably intrigued by the various ways in which rural people express their embeddedness in community and the landscape. De la Cadena quotes a bilingual Quechua-Spanish schoolteacher on the meaning of *ayllu*. He beautifully conveys the sensory experience of local rural identity, with which a great many humans all over the world could no doubt identify, but it takes a detached, modern individual to reflexively express – and to quote – that experience as a reality from which one has been estranged. Although de la Cadena repeatedly asserts that she is concerned with *relations* – for instance, when quoting Keith Basso or Roy Wagner – her approach to "in-*ayllu*" practices unfortunately seems to end up objectifying these experiential relations as if they were features of the people and the mountains whose interaction she seeks to explore. Although her efforts are sincere and painstaking, she is far from the first anthropologist for whom the phenomenological sense of place articulated by their rural hosts is inevitably lost in translation. To acknowledge and respect the web of relations that constitute such a sense of place is not equivalent to promoting, as literal statements about reality, the projections, attributions, and metaphors through which it is expressed.

Yet, this sometimes appears to be the ultimately untenable position pursued in *Earth Beings*, raising the question if the endorsement of animism is really an option. Are such perceptions viable, in the sense that they are being adopted by increasing numbers of young people? In the unlikely event that this is indeed the case, is it because there is money to be made in tourism? Can we perhaps expect a continued expansion of "Andean shamanism" catering to the projections of modern people attracted to embeddedness, spirituality, tradition, authenticity, and other qualities missing in their own lives? And if so, is not this commodified version of indigenous embeddedness something very different from Mariano Turpo's sacrificial *despachos* to Ausangate? How, in these complex transformations and imperfect translations, are we to understand the ontological status of the *tirakuna*?

De la Cadena's most remarkable sentences seem to suggest that the earth-beings invoked by the villagers of Pacchanta may indeed be real sources of agency. At one point she asserts that "our incapacity to be persuaded of their participation ... does not authorize the denial of their being" (ibid., 150). But then, how could we deny anything at all that someone might claim? What does denial mean? If we should not be permitted to deny something of which we cannot be persuaded, how would it at all be possible to have a conviction – for instance, a materialist worldview? De la Cadena at times interjects her own inability to perceive Ausangate as more than a piece of rock, but her disbelief is consistently suppressed and renounced as a personal shortcoming conditioned by modernity.

To refer to the relation of *runakuna* to *tirakuna* as a "belief," writes de la Cadena (ibid., 187–188), is to distance oneself from them, and Nazario Turpo was obviously not willing to think of his convictions regarding Ausangate in such terms. Regardless of worldview, to thus relativize one's own convictions would be to deconstruct the foundation on which one stands. Admittedly, the use of the word *beliefs* about others' convictions expresses the speaker's disbelief, but this is not necessarily tantamount to disrespect. To reject Cristóbal de Albornoz's and Alan García's intolerance of what they dismissed as "superstition" does not necessarily mean endorsing *runakuna* ontology. If the word *belief* were to be banished (as politically incorrect) from the anthropologist's vocabulary, the hyperrelativist implication would be that any conviction is as good as any other, and that anthropologists are prepared to share any conviction they encounter. However, as it will no doubt always be justified to distinguish between one's own convictions and those of others, an equivalent concept would soon have to take its place. As illustrated by the medieval tension between geocentric and heliocentric cosmologies, some beliefs are simply false. Although it is supremely legitimate to critically interrogate the political dimensions of the distinction between "our" rationality and "their" beliefs, the posthumanist campaign against modernity would be more subversive if it exposed the cultural and fetishistic foundations of modern rationality, rather than endorsing the validity of nonmodern animism.

To calibrate the human subject with earth-beings or other deities is a means of bracketing the self-reflexive ego-consciousness that is a diagnostic norm of modernity. For nonmodern people everywhere, it is a way of emphasizing the primacy of relations, rather than the objectified self. To reify and take literally the imagined, nonhuman interlocutor is a fallacy of misplaced concreteness. This becomes particularly evident when de la Cadena argues that Andean earth-beings are "part of the political process" (ibid., 93). If being targets of invocation qualifies them as political agents, then the same must be said for the Christian God – and crucifixes, altars, temples – through two millennia of European history. In fact, de la Cadena observes that indigenous Andean spirituality should be understood as a syncretic fusion of Christian and non-Christian practices.

A lingering question, then, is why the exclusion of *tirakuna* from public discourse is more deplorable than secularization in general?

The projection of anthropomorphic agency onto nonhuman entities is a way of relating to the world. To the extent that the concept of "belief" does not suffice to exhaust the substance of such relations, it is because beliefs generate real material consequences by significantly influencing human agency. But in contrast to nonsocietal events such as earthquakes, the reality of mountain deities presupposes subjective human perceptions. John Maynard Keynes distinguished between atomic and organic propositions, and the critical realists distinguish between transitive and intransitive aspects of reality. Posthumanists such as de la Cadena disregard the relevance of such distinctions between the objective and the subjective, but they would have helped her deal with the fact that the purposes of earth-beings impact on the material world only to the extent that they serve as concerns of human beings. If we are to believe that earth-beings have purposes and agency that is independent of human concerns, we need to see it demonstrated. Apparently, there were local voices suggesting that earth-beings were responsible for the tragic bus accident that ended Nazario Turpo's life (ibid., 19). De la Cadena writes that she doubts that this was the case, but, true to her narrative, she does not seem to rule out such metaphysical intervention *in principle*.

I sense a contradiction between de la Cadena's critical skepticism regarding the demand for Nazario Turpo's presence in the promotion of tourism, museum exhibits, and even individual politicians, on the one hand, and her reluctance to acknowledge the transformations that these new opportunities must have implied for his practice of "Andean shamanism," on the other. As elsewhere, the incorporation of indigenous spirituality into the modern market has generated competition between different practitioners, involving contestations of authenticity. Authenticity is the primary cultural capital commodified by the tourist industry, catering to disembedded moderns intrigued by the spirituality and embeddedness of the elusive native, yet it is inevitably undermined by this very commoditization. The *despachos* that Mariano had taught his son Nazario to direct toward Ausangate were a means of personal calibration, but when presented to tourists and museum curators they must have become something else: precisely the *objects* – objectified nature and objectified selves – so diagnostic of modernity. De la Cadena is, of course, quite aware of this dilemma but chooses – no doubt out of respect for her late friend Nazario – not to pursue the topic. Nevertheless, if contemporary anthropological culture theory can be used to obscure such discontinuities, it is tantamount to serving as an accomplice of neoliberalism.

The requirement of a modern secular individual, writes de la Cadena (ibid., 250), is the capacity to "distinguish cultural belief from rational knowledge." Because "Andean shamanism" is packaged for tourists and museums precisely as a set of "cultural beliefs," the implication is that people like Nazario must inevitably begin to experience themselves in such modern terms. But rather than pursue a consistent critique of how the market transforms culture, experience, and identity, de la Cadena at times seems to salute the "partial connections" and "equivocations" that characterize the interaction between local lifeworlds and the modern system. Interestingly, she also seems to endorse what she identifies among *runakuna* as a conflation of signifier and signified, a distinctly nonmodern phenomenon that may in fact have facilitated the transformation of Nazario's *despachos* into postmodern commodities. Contestations of authenticity appear more to have focused on the technicalities of ritual practice than on the modern preoccupation with degrees of correspondence between a person's surface and inner essence.

Much as *runakuna* are inclined to conflate signifier and signified,[2] they apparently tend not to distinguish between leader and follower, or representative and represented. Combined with their conflation of human and nonhuman members of *ayllu* – society and nature – this means that their elected leaders speak "from" the entire local lifeworld, including the *tirakuna*. This suggests a local version of the insight expressed by Roy Rappaport (1994) that humans are that part of the biosphere that can reflect over itself. But being immersed in totalities that span the divide between society and nature does not absolve humans from their unique responsibility to speak and care for such totalities. When elected as *personero*, Mariano Turpo was expected not only to speak on behalf of the *ayllu* but also to be the *ayllu* speaking. If this is anthropocentrism, it is not hierarchical in the sense of conceiving humans as superior to the nonhumans with which they identify.

Whereas "in-*ayllu*" leaders are identified with their constituencies, more distant leaders such as *hacendados*, police chiefs, lawyers, and urban politicians are frequently portrayed as corrupt and oppressive. Their capacity to exert sovereign power over *runakuna* at will,

---

[2] But in the Museum of the American Indian in Washington, DC, Nazario was prompted to distinguish between a picture of Ausangate and the actual mountain.

encapsulated in the concept of *munayniyuq*, is conceptually assimilated with the power of *tirakuna*. While attempts are made to placate these powerful agents, as with gifts of sheep or *despachos*, both are ultimately beyond control. When such parallels between the supplication of human and nonhuman masters are acknowledged, the Andean invocation of earth-beings invites a rather conventional interpretation quite familiar to students of magic, ritual, and religion (cf. Sillar 2009). However, this does not deflate the value of de la Cadena's close scrutiny of its personal and political dimensions. Although her posthumanist conclusions can be contested, she raises important questions about the role of ethnography and anthropology in a world where cultural convictions and identities cannot avoid being transformed by the self-reflexivity of neoliberal modernity.

Nevertheless, an insidious political aspect of the indigenous condition is left out of the picture in de la Cadena's account. Neoliberal market ideology promotes the universal solubility of exchange-values – commodifying Andean "shamans" and mountains alike – but the insidiousness of neoliberal logic is that the superficially democratic approach of multiculturalism tends to render the specific cultural content of human lives arbitrary, as long as people adhere to the premises of the market. If cultural identity is understood as something more existentially profound than adopting a fashion, enjoying folk music, or choosing a restaurant, a decolonial anthropology would need to focus on how such meaning is dissolved by neoliberalism. Rather than endorsing this dissolution of meaning, a truly decolonial anthropology must struggle to expose the cultural foundations of the market. Neoliberal concessions to diversity are instrumental to market logic at the submerged level where this logic is rarely questioned. An anthropology that truly transcends coloniality must seriously subject modernity – and the market – to cultural critique – not simply by attempting to resurrect that which modernity has supplanted but instead by identifying its own cultural logic.

It can also be argued that de la Cadena mistranslates relationism. To talk about people's *relations* to mountains is not equivalent to talking about those people or those mountains. Relations should not be objectified into features of people or landscapes. It is finally unclear if de la Cadena thinks of earth-beings as denoting human-environmental relations or features of mountains. Does Ausangate have agency? Can earth-beings purposefully cause bus accidents? De la Cadena's ambiguous translation of *runakuna* relationism risks converging with the touristic reification of rural Peruvians into "shamans" and their mountains into "earth-beings" – precisely the objectified selves and objectified landscapes that are so diagnostic of modernity.

I would not follow de la Cadena in attributing animateness to mountains or endorsing attempts to communicate with them – no more than I believe that prayers will stop climate change. While indigenous claims about the sanctity of a mountain may at times serve to protect it from extractivism (cf. Hornborg 1994),[3] to sympathize with the political effects of those claims is not equivalent to subscribing to indigenous ontology. De la Cadena's book raises questions about the role of secularized anthropologists in environmental struggles such as the movement to save the mountain Ausangate from a mining project. Should anthropologists be prepared to adopt any conviction they encounter, or is skepticism or even denial at times appropriate? Does the Christian God – and associated artifacts – have political agency? Is the exclusion of earth-beings from legal discourse more deplorable than secularization in general? But the most significant omission in de la Cadena's account is the role of money and market logic in commodifying cultural identities, worldviews, ritual objects, and geological formations alike. While championing what she perceives as the nonmodern, she thus fails to address the problems of modernity.

Relationism as Revelation or Prescription?

Tim Ingold's work, while solidly grounded in anthropology, builds bridges to literature in philosophy, biology, art, architecture, and other fields rarely explored by anthropologists. He, too, wants to challenge modern ontologies, often by contrasting them with those of nonmodern people, but proposes to say something about what human life universally *is*, rather than suggesting that different groups of people live in different worlds. For him, the European concept of animism is a misunderstanding of the propensity of nonmodern people to *relate* to the various components of their environment, whether a modern European would classify

---

[3] In my study of how indigenous Mi'kmaq saved a mountain in Nova Scotia from being turned into a granite quarry (Hornborg 1994), I emphasized how their concept of sacredness served to disarm the modernist discourse founded on interchangeability, quantification, and calculation. I also argued that these opposite outlooks express a social polarity regarding contrasting modes of identity formation – between persons constituted in terms of concrete and specific points of reference (noninterchangeable places, people, and things), on the one hand, versus more abstract and general categories of spaces, social relations, and commodities, on the other.

those entities as biologically alive. Rather than bounded objects, Ingold shows, organisms are bundles of entangled relations. To him, this is not just a matter of how life is perceived but a factual condition.

Ingold's work poses a central conundrum to anthropology because it amounts to an ontological critique of the modern worldview but without identifying power relations as a crucial aspect of that which is criticized. It is no coincidence, and no less tantalizing, that Ingold (1988, 1997, 2000) has been the leading proponent of an "anthropology of technology" (see Chapter 6), yet without addressing the political dimension of modern technology. As I hope to have made clear in earlier chapters, I do not emphasize the political aspects of technology simply because of a compulsion to see all human phenomena from a political perspective, but because power inequalities are inextricably *constitutive* of the machine. The relation between power inequalities and technology is an analytical and intellectual challenge, rather than just a matter of ideological perspective. Power is a central analytical issue in social science. To address it is not to "politicize" essentially nonpolitical concerns – as if social organization could ever be viewed as politically neutral – but to penetrate the very essence of the social. In other words, to view technology as a phenomenon that can be excised from relations of power inequalities is to fail to grasp its societal rationale: its requisites, its consequences, and the foundations of its very existence. To briefly reiterate the argument in this book, fossil energy technologies are not simply a replacement of technologies for exploiting solar energy – including manual labor – but a strategy for *expanding* such exploitation. To unravel their societal and ecological consequences is fundamentally aligned with the general goal of political ecology to show how environmental issues and power relations are intertwined. In the remainder of this chapter I will try to show how Ingold's critique of modernity can be harnessed to political ecology.

Ingold's (2018) brief but useful autobiography "From Science to Art and Back Again" outlines the progression of his concerns from biology through cultural ecology and Structural Marxism to a phenomenologically inspired, monistic *relationism* evoking affinities with art. He celebrates the wondrous diversity of biological and material forms continuously generated in the myriad relations in which living things are immersed. As he suggests, this childlike fascination with the wonders of the natural world is no doubt in part the heritage of past generations of biologists, whose sense of awe is combined with gratitude. Precisely because I share his biophilia (Wilson 1984), however, I am curious about what he has to say about contemporary *threats* to global biodiversity.

Considering the state of the planet today (cf. Kolbert 2014), I would expect his indignation to be proportional to his sense of wonder and gratitude. In what follows, I will thus focus on the relation between his extensive insights regarding the processes by which living forms are generated, on the one hand, and his conspicuous silence on the processes by which they are destroyed, on the other. While the former concern necessarily emphasizes domains of local experience, the latter must address global political ecology.

The boundless, holistic aspirations of anthropology open doors in both these directions. Ingold has thoroughly explored and charted the sensory, perceptual engagement of humans with their immediate environment, but hesitated to venture into the abstract territory of global political economy. Although the two concerns require distinct conceptual tools, they are not unrelated. It can be argued that the logic of global markets and the concepts of mainstream economics are precisely what has obscured that which Ingold has been struggling to articulate. Anthropologists are generally receptive to both kinds of observations: the exploration of the experience-near, extralinguistic involvement of humans in their material surroundings, and the analysis of the discursive and political frameworks that *condition* such involvement. To a significant extent, the latter conditions have for more than two centuries increasingly constrained most humans from experiencing the world in the way that Ingold advocates. Those conditions have been variously called modernity, capitalism, the market, industrialism, and so on, but Ingold only occasionally and briefly confirms that these are indeed the targets of his critique. Many critical anthropologists have chosen to direct their discontents explicitly at such abstractions, and to theoretically dissect their historical, political, cultural, and ontological foundations, but Ingold's mission has been to familiarize us more directly with the world that we have lost, or rather with the world as we could have experienced it.

This project is in itself a formidable undertaking. It has convinced many of us that the lenses and cognitive filters through which we moderns have learned to perceive our environment may have distorted our experience of being human. But such an observation raises several problematic questions for a discipline accustomed to celebrating cultural relativism. When Ingold asserts that living organisms should be perceived as "bundles of lines" rather than "blobs" (2018: 216), is he suggesting that the former view is more correct, and the latter mistaken? How would he account for the fact that most people on Earth now adhere to the latter perception? What is the source of his alternative insight? Does it derive from his experiences of ethnographic fieldwork among the Saami or from a practice of handicraft or art? If a majority of humans could be persuaded to reconceptualize organisms as lines rather than blobs, would it make a difference to the future of the world?

I believe that all these questions can be given answers that in no way detract from the validity of Ingold's perspective, but I am frustrated by his tendency to ignore them. Whether a product of our sensory constitution or of the Enlightenment, the reification of organisms conceived as bounded "blobs" rather than interpenetrating bundles of relations can be challenged in two quite different ways: by emphasizing either their *experiential* relations (as does Ingold) or the *material* relations of exchange through which they exist and are reproduced. Given his background in biology and ecology, Ingold is naturally aware that organisms should not be excised from either of these aspects of their fields of relations, but his more or less exclusive focus on the former tends to obscure the crucial fact that the semiotic and the material aspects of ecological relations are mutually constitutive. The most significant implication of Jakob von Uexküll's (2010) *ecosemiotic* perspective is that the interaction and coevolution of myriad subjectivities (*Umwelten*) is as fundamental to the constitution of ecosystems as the flows of matter and energy to which the quantitative methods of modern natural science are confined. Ecology is essentially about relations of interpenetration that are both semiotic and material, but the modern science of ecology only recognizes the latter.

Uexküll clearly recognized an analytical distinction between an organism's subjective and perceptual *Umwelt,* on the one hand, and its objective biochemical conditions and processes, on the other. This distinction is cognate to the emic/etic distinction that was so prevalent in the cultural ecology that Ingold abandoned in the 1970s. Since then, Ingold has experimented with several approaches to integrating both the subjective and the objective and the social and the natural. In the early 1980s, he reminds us, he followed Structural Marxism in rejecting the "vulgar materialism" of cultural ecology, for which the distinction between the subjective (emic) and objective (etic) was equivalent to the *determination* of the subjective by the objective – culture by ecology. Following Baudrillard, Marshall Sahlins (1976) turned this model upside down by showing that culture instead unfolded according to its own semiotic logic, and Structural Marxists such as Maurice Godelier (1978) and Jonathan Friedman (1974) emphasized the recurrently contradictory relation between the relatively autonomous levels of society and nature. In its essentials, the latter approach remains compatible with the contemporary recognition that globalized capitalism is unsustainable. For many anthropologists and other social scientists, however, the aspiration to establish the character of the relation between an abstract society and an abstract nature ultimately led to a desire to dissolve the boundary between them. For Tim Ingold, Bruno Latour, and many others, the very categories "society" and "nature" emerged as obsolete (cf. Latour 1993a; Descola and Pálsson 1996). As Ingold recalls, it was his failure to "hive off the social from the ecological" that drove him to sympathize with artists "struggling to break down the boundaries between the human and the non-human" (2018: 225).

Ingold's intellectual autobiography (2018) reflects on the emergence of such efforts in a way that makes it possible to understand the logical trajectories of these transformations of anthropological inquiry. Whereas the 1970s saw a widespread rejection of materialism – the notion that the quantifiable and physical is causally prior to the subjective and experiential – the subsequent swing of the anthropological pendulum appears to have thrown the baby out with the bathwater. In Ingold's case, this is evident in his inattention to precisely those quantifiable and physical aspects of organisms and societies that were overly prominent in the agenda of cultural ecology. We all know that flows of energy and materials continue to be essential to both social and ecological systems, yet they are almost completely invisible in contemporary anthropological deliberations on society and nature. But being denied determinacy should not be tantamount to disappearance. To truly integrate the social and the natural would be to show how flows of energy and materials are *used* to generate social organization – that is, how human societal relations enlist nonhuman nature in the service of their own logic. That logic, I would emphasize, generally hinges on power and inequalities.

When Ingold abandoned cultural ecology, he recalls being persuaded by Sahlins's culturalist arguments and Godelier's Structural Marxism, but he does not mention the school of political ecology. Yet Eric Wolf's (1972) challenge to cultural ecology founded the political ecology approach in anthropology, and the wide adoption of the latter reflected a widespread discontent with the local and adaptationist focus of the former school (Watts 2015). Although no less anchored in local case studies, a hallmark of political ecology is its constant awareness of global political economy. Eric Wolf (1982) taught anthropologists to rethink cultural and ecological processes on all continents as consequences of economic and political processes at the level of the world-system. Political ecology showed us how to transcend the local and adaptationist outlook of cultural ecology but without abandoning materialism. It reframed the relation between society and nature by recognizing that society over the past five centuries has been nothing less than global.

Ingold's concern with the experiential details of microlevel human engagement with the nonhuman environment seems a project far removed from tracing the trajectories of the world-system, but they are not unconnected. Although it would be unreasonable to request a single scholar to explore both these disparate aspects of human-environmental relations, the question arises how Ingold conceives of the relation between world-system integration, on the one hand, and the historical transformations of experience, on the other. His recurrent focus on nonmodern, nonindustrial modes of engaging the environment suggests an implicit critique of industrial capitalism and the world market that spawned it, but rather than explicitly deploring the extensive shift from tools to machines – and from practical skill to abstract blueprints – he seems to ask us to rediscover a mode of experiencing the world that modernity has obscured from view. However, this immediately raises the question if our loss of relatedness can be distinguished from real-world processes of reification and objectification, and if an effort to rekindle relationism can be pursued independently of a political confrontation with the social system that continues to transform relations into things. Like his predecessors in phenomenology, psychology, anthropology, and art, Ingold recognizes the profoundly meaningful realms of experience of which capitalist modernity tends to deprive us. He eloquently persuades us of what we have lost, but not to stop at nostalgia, his championing of relationism deserves to be connected to questions of power.

Power relations are based on convictions about the character of the world. On reading his various texts, I ask myself whether Ingold subscribes to a realist or constructivist ontology. Are relationist accounts of human activity equally applicable to all social contexts, whether nonmodern or modern, or does the modern treatment of the world as a collection of "blobs" make the world – and humans – more bloblike? Ingold is explicitly critical of the ontological turn in anthropology, emphatically affirming that the world is *one*, but there are moments when his relationism verges on a relativist or even solipsist ontology. Does not his denial of an objective reality ultimately imply an assertion of ontological diversity? An example of such ontological ambiguity is his suggestion, inspired by Goethe, that "the sun we perceive in the sky, and that lights the world of our experience, can *exist* only through its essential correspondence with the eye" (Ingold 2015: 99; emphasis added). While I readily agree that, for social scientists, the relation between the knower and the known recursively contributes to shaping both, I cannot accept that the same observation applies to astronomy. There are what the critical realists call *intransitive* aspects of reality that exist independently of human cognition, and that require modifications of a general relationist approach to knowledge. While knowledge is always a relationship, some objects of knowledge remain completely resistant to the way in which they are conceived.

As already indicated, a central issue is what we mean by "society" and "social." When Ingold in the 1970s and 1980s struggled to reconcile the dualist understanding of humans as social *and* ecological beings – persons *and* organisms – his concepts of "social" appears to be based on the social organization of hunter-gatherers, referring to the "relations of food sharing and the division of labour between men and women" (2018: 220). In this view, human society is a matrix of "personal" relations. Ingold realized that person and organism in hunter-gatherer societies were indissolubly one and the same, and that the implications required a "completely different kind of thinking, one that starts not from populations of individuals but from fields of relations" (2018: 221). But the social and ecological matrices of relations in which modern people are immersed are very different from those of hunter-gatherers. They are generally not local and personal but global and impersonal. They also involve the distribution of food and labor, but not so much through interpersonal sharing and collaboration as in the form of abstract commodity markets and international trade. The crucial difference between modern people and hunter-gatherers is thus that, for the former, the matrices of relations that constitute the organism and the person do *not* coincide. Physically and objectively, the molecules that compose our bodies tend to derive from the far-flung corners of the planet,[4] but personally and subjectively, most of us remain embedded within a much more limited social matrix of relations. Even if modernity has radically transformed the scope of our social contexts, the scale of our economic and ecological reach far exceeds our subjective and experiential reach as persons.

Moreover, a political ecology approach cannot be content with a mycological or fungal metaphor for society, even if conceived as a global matrix. If the world-system has rhizomatic features, its myriad lines of

---

[4] This difference between the constitution of nonmodern and modern people is captured in Raymond F. Dasmann's distinction between "ecosystem people" and "biosphere people."

relations must be recognized not merely as an invalidation of bloblike representations of nations or cities, but as conduits of *asymmetric* flows. While mutually constitutive of the nodes that they connect – generating and reinforcing core-periphery relations – those connections are not politically neutral but sources and means of capital accumulation. Fungal metaphors tend to obscure such power asymmetries by emphasizing mutuality and connectivity at the expenses of inequality and exploitation. It is thus tantalizing to discover, interspersed in Ingold's (2018) text, fragments of an explicit critique of capitalist modernity. He might, for instance, have elaborated his assertion that anthropologists may "sidle up to power and chip away at its pretensions" (2018: 216). What does "power" mean here? Perhaps he refers to "the colossal expansion, over the last four decades, of globalization and the political economy of neoliberalism" (2018: 217)? But what does he mean when he proposes that anthropology "can help pave the way for sustainable futures" (2018: 225)? If the unsustainability of contemporary life is inherent in the outlook and discourse of neoliberalism, how can anthropology challenge it? Ingold's contempt for the commodification of science and for "the neoliberal economy of knowledge" evokes a contradiction between a "global scientific elite ... in collusion with the corporations it serves" and an "increasingly impoverished" world population (2018: 226). I wholeheartedly share his campaign for care, responsibility, and truth, but to denounce the scientists is merely to chip away at the tip of the iceberg. While Ingold's anthropology persuasively reveals the dimensions of experience of which we have been deprived, the comparative horizons of anthropology can also be used to examine the submerged contours of the iceberg on which is founded not only science but also neoliberal power, unsustainability, and global impoverishment. At the root of all these evils are the very ideas of general purpose money, the global market, and the progress of its technological offspring since the Industrial Revolution.

Ingold's (2000, chs. 15–16) own analysis of the phenomenon of modern technology identifies its externality to the human body as a significant historical rupture, but its requisites are no less economic than cognitive or experiential. In his own words, "[T]echnical relations have become progressively disembedded from social relations, leading eventually to the modern institutional separation of technology and society" (ibid., 321–322). As I have argued in previous chapters, we need to rethink the nature-society distinction to realize that machines may be instruments not just for putting nature to work but also for putting other segments of global society to work. The Industrial Revolution – and the history of technology ever after – certainly required engineering science, but it has been no less dependent on the asymmetric resource flows of the world-system. Because we tend to think of modern technology as revealed nature, ontologically sequestered from the world market that is its requisite, the harnessing of inorganic energy in the first combustion engines has yet to be understood as a global societal event. Our conventional separation of economics (as the study of social exchange sequestered from nature) and engineering (as contingent on market prices, yet conceived as the revelation of nature) tends to obscure the increasingly obvious fact that economic growth and technological progress are inextricably connected euphemisms for exploitation. For more than two centuries, globalized market trade conceived as politically and morally neutral exchange has entailed the displacement of work and environmental loads from wealthier to poorer parts of the world-system. From the colonial slave plantations that supplied British factories with cotton fiber to the sugarcane plantations that now provide European cars with ethanol fuel, technological progress in the core has been founded on the appropriation of human time and natural space in the periphery.

Ingold's interest in identifying the specific characteristics of the phenomenon of modern technology is not incidental, as it defines the difference between the living and the nonliving. It ultimately also bridges the analytical chasm between phenomenology and political ecology. As I suggested in a review of Ingold's book *The Perception of the Environment* (Hornborg 2002), technology has provided a master trope for the distorted views of both culture and biology that Ingold so persistently criticizes. Machines are counterfeit organisms – inanimate replacements of living processes such as human laborers and draft animals. Marx's notion of "dead labor" captures their zombielike character but does not sufficiently acknowledge the extent to which they depend on asymmetric social transfers of energy and other biophysical resources. Machines are strategies for harnessing physical forces and substances in nature to reproduce power inequalities in society but are couched in the politically neutral idioms of economics and engineering. They are paradigmatically *socionatural* phenomena but appear to our consciousness – viewed through the twin filters of economics and engineering – as sequestered from society and politics. Our civilization is committed not only to the illusion of producing machines that are more efficient at harnessing solar energy than living beings but also to obscuring the exploitative foundations of technological progress. Ingold's fascination with the experience of being a living being has led him to identify what distinguishes a machine from an organism, but his insights deserve to be elaborated into an even more profound – and subversive – philosophy of technology. More than ever, given current deliberations on cyborgs and artificial intelligence, we need to maintain a precise distinction between persons and things. But regarding precisely this issue, I find Ingold's position ambiguous. He explicitly asserts that for him, "[T]here are no objects" (Ingold 2018: 224). But surely machines *are* objects? To the extent that we downplay the distinction between sentience and nonsentience (subjects and objects), we risk succumbing (like Latour and his followers) to fetishism – the attribution of animateness to nonliving things.

In the sense that they are inanimate and nonsentient, machines are definitely objects. But if we follow Ingold in opposing objectivism to relationism, we must conclude that machines, too, are "bundles of relations." The functioning of a tractor is as dependent on inputs of fuel energy as the functioning of an organism is on inputs of nutrients. There are, as previously pointed out, two ways in which an entity is contingent on its relational context: the phenomenological sense extensively explored by Ingold and applicable only to living beings, and the material or metabolic sense that living beings share with machines. It is the sequestration of economics and engineering that permits us to perceive machines as "blobs" that exist independently of global exchange relations. This illustrates how disciplinary fragmentation fosters reification and fetishism. To promote a relationist perspective on living organisms as simultaneously sentient and material forms would require the kind of interdisciplinary synthesis of phenomenology, anthropology, and biology that Ingold has been articulating for decades, but complemented with attention to the flows of energy and matter in ecosystems that used to preoccupy cultural ecology. To promote a relationist perspective on technology, however, would require a synthesis of economic history, ecological economics, and political ecology. It would require us to ask, whenever deliberating on a new invention, to what extent it is merely a way of putting nature to work, and to what extent it is a way of putting other segments of global society to work. In other words, it would require us to acknowledge that technologies are not just revelations of intrinsic properties of nature, but *socio*natural phenomena that reflect the metabolic structures of world-systems.

Ingold (2018: 217) suggests that the modern abandonment of Goethean ideals of close sensory familiarity with the objects of a scientist's attention should be seen as a surrender to neoliberalism, and that the best we can do to counter the commodification of science is to turn to art. While I am persuaded that some aspects of human-environmental relations are more effectively communicated through art than through prose, the ideal of close sensory familiarity or correspondence with an object exposes the limitations of phenomenology because no degree of sensory familiarity with a machine will reveal the global field of relations of which it is a manifestation. A commitment to phenomenology can thus serve to obscure macrolevel relations such as the structures of global political economy. To challenge "globalisation and the political economy of neoliberalism" (ibid.), which Ingold explicitly deplores, requires that we transcend Goethean ideals and address the logic of abstract systems such as the world market.

I believe that the dilemma that has propelled Ingold's (2018) pendulum movement ("from science to art and back again") ultimately boils down to the insidious way in which we tend to conceive of the distinction between the semiotic and the material – the subjective and the objective – as a *causal* relation. Ingold was as justified in abandoning the materialism of cultural ecology as he was in not simply adopting the opposite view that cultural representations are arbitrary semiotic systems that impose their autonomous logic on the organic and physical. His struggles to find a compromise between Rappaport and Sahlins have consistently sought to integrate the material and the experiential without positing either as somehow prior to the other, but as previously observed his phenomenological solution tends to leave some important questions unanswered. The experiential or subjective aspect of human-environmental relations necessarily implicates the continuous formation of cultural and discursive frameworks for conceptualizing and perceiving the world. Ontologies are produced not merely through individual engagement with the physical environment but also simultaneously through social processes of narration, categorization, and emulation. Such social processes of meaning formation define what we too unreflectingly tend to refer to as "cultural construction." Their tacit and sensory dimensions certainly deserve emphasis, but this should not entail completely jettisoning the significance of the linguistic and discursive. Whether hunter-gatherers articulating their experiences of the nonhuman environment or nineteenth-century Britons consolidating their outlook on industrialism and world trade, humans continuously generate ontologies that have material as well as sensory repercussions. Such ontologies are neither adaptations to the environment nor semiotic idiosyncrasies blindly imposing their autonomous trajectories on physical reality, but collectively negotiated human modes of relating to the world, responding to its materiality, and shaping human activity. Viewed as a process of continuous mediation of human-environmental relations, it would be as misleading to think of culture as determined by the environment as it would be to think of it as free to pursue its autonomous logic. As Ingold indicates, culture is not an add-on to biology, but the human version of a universal semiotic capacity essential to all life. It is indissolubly part of our practical and material engagement with our environment.

Ingold's interventions ultimately highlight the fact that the relation between semiotic and material aspects of human-environmental relations is fundamentally transformed with globalization. In the modern world, the sphere of sensory engagement with artifacts and organisms does not coincide with the social conditions through which those artifacts and organisms are produced. Ingold's long-standing efforts to transcend dichotomies such as "economic versus ecological, social versus natural, person versus individual" (2018: 221) are highly justified in the context of local groups of hunter-gatherers but are not as applicable in the modern world-system. We have observed that, rather than being indissolubly one, the material derivation of the modern human organism tends to be geographically much more diffuse than that individual's constitution as a person. In similar ways, the operation of the world economy is disembedded from the destinies of particular ecosystems. The trajectories of our artifacts (such as money and technologies) and discourses (such as economics and engineering) have been detached from the sensory experience and the cognitive reach of individual humans. In this sense, modern ontology may seem free to pursue its own logic but is continuously confronted with the material constraints of the biosphere within which it unfolds, as evidenced by our growing concerns with sustainability. The illusory emancipation from ecological constraints that pervades the modern outlook is a cognitive corollary of the physical displacement of environmental burdens to other continents. At the discursive level, at least, the illusory character of this outlook is now widely acknowledged. At the practical level, a serious effort to implement policies for sustainability that transcend mere rhetoric would need to rewrite the rules of the global economic game so that it is no longer geared to neocolonial, asymmetric transfers of resources to world-system cores.

As a leading environmental anthropologist, Ingold can be expected to have substantial things to say on sustainability and environmental justice. Yet, when he mentions the Anthropocene (2018: 225), he suggests instead turning to art as a source of humility and "radical ecological awareness."

He does not seem concerned with how issues of global justice and sustainability are intertwined. Even when he has discussed the weather (Ingold 2015: chs. 11–15), I have found little to persuade me that phenomenology can help us combat climate change. Although impressed with his eloquence and the scope of his erudition, I am left wondering how art and phenomenology can challenge the abstract logic of economics, engineering, and globalized capitalism.

I realize, of course, that these are not issues easily addressed on the basis of the premises that run through Ingold's rich and interdisciplinary project. His attention has focused on the kind of sensory relationism that can be applied to the immediate human engagement with the nonhuman, not on revealing that even machines, in a cognate way, are "bundles of lines" rather than "blobs." But, as I indicated at the outset, his efforts to resurrect nonmodern modes of existence and perception – and to celebrate the diversity of biological and material forms – conveys an implicit but powerful critique of the forces that threaten such modes of being and such diversity. This is why I would encourage him to explicitly address how our modern ontology tends to undermine life, as an indissolubly semiotic and material phenomenon. The germ of such a line of inquiry can be found in his insight that the historical emergence of a concept of disembedded technology is completely parallel to anthropological analyses of the emergence of the disembedded economy (Ingold 1997: 108). This is to say that Ellul's (1964) insights on technology are cognate – and in fact concomitant – to Polanyi's (1944) on economy. Here is the very essence of modernity, the global and decontextualized abstractions of which Ingold challenges with his pervasive focus on the local, concrete, and embedded. Given his rich understanding of both ecology and the conditions of human experience, Ingold is well placed to elaborate a critique of the simultaneously semiotic and material consequences of practices and discourses that disembed artifacts, landscapes, and not least *people* from bundles of relations that are local enough to be accessible to experience.

In this chapter, I have discussed two different ways of challenging modern ontology. One views the occurrence of animism as evidence of the multiplicity of realities and the contextual validity of imputing agency to what we would categorize as objects; the other sees animism as a reflection of the actual, relationist essence of reality. One interprets narratives of relatedness as accounts of exotic worlds, the other as glimpses of the world as it might be experienced. Neither of these approaches, unfortunately, provides a decisive challenge to the modern worldview or to modern power structures. The former attempts to mobilize an ontology that has proven incapable of resisting modernity, the latter fails to address the root causes of our modern alienation. Neither of them considers what *money* does to our experience of ourselves and our relations. Although Ingold's early struggles to establish an "anthropology of technology" held much promise, nor does either of these efforts to defamiliarize modern ontology prompt us to fundamentally revise our image of the machine. An ontological turn that might help us handle the Anthropocene would not be preoccupied with scrutinizing the worldviews of people who have no responsibility whatsoever for climate change, but to turn the anthropological gaze back at those who do. Its ontologically most disturbing discovery would be that neither money nor machines, nor even energy, are what we have long thought they were.

13

Conclusions and Possibilities

In this concluding chapter, I address the implications of my argument throughout this book that many destructive aspects of the contemporary global economy are consequences of the use of general purpose money to organize social and human-environmental relations, and that the political ideals of sustainability, justice, and resilience will only be feasible if money is redesigned. The argument is based on the conviction that human artifacts such as money play a crucial role in organizing society, and that closer attention should be paid to the design and logic of such artifacts, rather than devoting disproportionate intellectual energy to theorizing and meticulously tracing their complex repercussions in terms of patterns of human behavior, as if those patterns were amenable to regulation without redesigning the artifacts that generate them. What is generally referred to as capitalism, I maintain, is the aggregate logic of human decisions about the management of money. Visions of a postcapitalist society using conventional money is thus a contradiction in terms. The chapter sketches a possible redesign of money based on the idea that each country establishes a complementary currency for local use only, which is distributed to all its residents as a basic income. The distinction between two separate spheres of exchange would insulate local sustainability and resilience from the deleterious effects of globalization and financial speculation. To indicate that the suggestion is not as unrealistic as it may seem at first sight, I briefly and provisionally respond to some of the many questions raised by the proposal.

Although soon reduced to a heterodox and marginalized position,[1] the critique of growth launched in the early 1970s has continued to challenge mainstream dogmas of economics and policy for more than four decades. A wide range of debates have raged on what exactly is the problem of sustainability, whether the design of taxes and subsidies, the choice of energy sources, the internal contradictions of capitalism, or even the biological essence of the human species. Considering the centrality of the focus on the economy, it is remarkable how little attention has been paid to the phenomenon of money. Mainstream economists certainly do not question money, but neither do most Marxists or ecological economists. Even Georgescu-Roegen (1971), who is recognized as the origin of both ecological economics and the proposal for degrowth, wrote a 457-page book on the contradiction between economics and thermodynamics without once asking if the cultural convention we know as money might in fact be the elephant in the room.

The use of money in a particular context reflects the ideas people in that context have about the interchangeability or commensurability of various goods and services. Different kinds of money have been used in different social and cultural contexts for thousands of years, but the extent to which goods and services became interchangeable on the market in nineteenth-century Europe was unprecedented. To say that land and labor were extensively commoditized in this period is tantamount to saying that they became exchangeable for money. In premodern societies, monetized exchange was generally limited to long-distance trade in preciosities, while most basic needs were provided through socially embedded relations of reciprocity and redistribution, but in nineteenth-century Europe money became a medium to obtain all kinds of goods and services, including food and other everyday needs. This is what economic anthropologists mean by general purpose money, as distinct from different kinds of nonmodern special purpose money, which reflect more limited principles of commensurability within distinct spheres of exchange.

As illustrated by the assumptions and prescriptions of mainstream economics, there is an inherent logic in general purpose money that

---

[1] Fifteen years after the UN Stockholm Conference on Environment and Development in 1972, much of the public hesitation about economic growth had been reined in by the neoliberal and ecomodernist discourse on "sustainable development" promulgated by the Brundtland Report (World Commission on Environment and Development 1987), which has been hegemonic ever since.

promotes an ambiguous form of "efficiency" that implies maximum exploitation of global differences in wage levels, land prices, and environmental legislation, which means encouraging long-distance transports, economic and environmental inequalities, and a pervasive indifference with regard to the distant social and ecological consequences of consumption. Although we tend to deplore such negative aspects of globalization, we may have to concede that they are inevitable implications of an economy organized in terms of general purpose money. Given the general imperative of minimizing costs and maximizing income defined by the logic of such money, mainstream economists may be right in that there is no alternative – within their field of vision. From their perspective, it is only to be expected that nations scramble to engage in a global "land grabbing" to ensure future access to key resources (Seaquist, Johansson, and Nicholas 2014). Even the heterodox economists and sociologists who acknowledge the inexorability and deleteriousness of "ecologically unequal exchange" as inherent in modern world trade have no remedies to offer (Frey, Gellert, and Dahms 2019). The aim of this chapter, however, is to sketch the operation of a different kind of currency system, which would promote a noncapitalist definition of efficiency, encouraging local resource management and frugality, social and ecological resilience, food security, and community. Whereas general purpose money tends to generate economic globalization, the proposal for a complementary, special purpose currency presented here is offered as a way of *relocalizing* the bulk of economic interaction.

If, in the early 1970s, many people realized that there were several problems with a societal focus on economic growth and productivism, few acknowledged that the imperative to pursue growth is inherent in the phenomenon of general purpose money. The imperative of minimizing costs and maximizing income, and the myriad practices, regulations, and institutions through which such strategies are organized, reflects the logical repercussions of the very idea of a universal solvent making all needs commensurable. Few people tend to reflect over general purpose money as a cultural peculiarity to which there are alternatives. Economic anthropologists, however, will recall Paul Bohannan's (1955) account of the "multicentric" economy of the Tiv of Nigeria, which exemplifies the widespread occurrence among premodern peoples of a hierarchy of economic values, in which different categories of goods and services are not readily exchanged for other categories of values. Among the Tiv, for instance, there was hesitation about exchanging labor or prestige goods for food. Regardless of the particular idiosyncrasies of such distinctions, the principle that they express – the assertion of *limits* to commensurability – is a radical challenge to the market ideology that has since transformed the economy of the Tiv and so many other indigenous peoples throughout the world. Although the idea is anathema to mainstream economics, the argument in this chapter is that it may be a key to alleviating global problems of increasing inequalities and declining sustainability.

I will go even further by suggesting that it is precisely in their inability to think beyond conventional money that heterodox approaches such as Marxism and ecological economics fail to offer feasible alternatives to business as usual. The exploitative dynamics of capitalism and the pervasive, asymmetric transfers of environmental resources are expressions of the logic of general purpose money, which assumes and encourages the exchange of industrial products for increasing amounts of the resources spent in producing them. The metric offered for measuring both "surplus labor value" (Emmanuel 1972) and the "value of ecosystem services" (Costanza et al. 1997) is always money. In both cases, the envisaged inadequacy of compensation is a logical consequence of the systematic struggle of market actors to keep costs lower than incomes. When such "efficiency" is pursued not at the local level, where Adam Smith identified its benefits, but in a globalized economy where fossil fuels have minimized transport costs, the scope for power differences, polarization, and exploitation – and thus the collateral damage of "efficiency" – is vastly greater. "Efficiency," in fact, is inverted into its opposite. As long as we subscribe to the assumption of general purpose money as the medium of exchange organizing human societies, exploitation and environmental degradation are inevitable implications of production processes. Although he did not draw this conclusion himself, it is solidly founded on the interdisciplinary insights of Nicholas Georgescu-Roegen (1971), who famously showed that the production of exchange-value simultaneously increases biophysical entropy.

The period after the 1971 abandonment of a gold standard has revealed the long-term trajectory of general purpose money. Since the introduction in Europe of paper money in the fourteenth century, and the first major financial crisis in Florence in 1343 (Weatherford 1997: 78), the risks of detaching money from finite, material parameters have repeatedly become evident. In the 628 years between 1343 and 1971, the aggregate consequences of states, banks, and market actors maximizing their monetary assets have included long-term tendencies toward increasing economic inequalities and environmental degradation, punctuated by points of extensive monetary devaluation, illustrating the fundamental vulnerability of states, financial institutes, and people in general. Although a great number of policy suggestions have been presented to attempt to remedy such recurrent tendencies and events, mainstream proposals do not critically scrutinize the inherent features of money. Nor, as we have observed, do most proposals from heterodox camps such as Marxism and ecological economics. This chapter considers what such a critical scrutiny would entail, and some possible conclusions. It does not belittle the technological and societal accomplishments in Europe since the fourteenth century – most of which have been connected to entrepreneurship in the context of a capitalist economy – but suggests that current concerns with climate change and financial crises offer a historical moment for reflection on how the operation of the global economy might be reorganized in the interests of global sustainability, justice, and financial resilience.

The theory of comparative advantage that underlies the promotion of free trade is valid only when the trade partners are obliged to comply with common legal regulations: when this is not the case, low wages and feeble environmental legislation may constitute competitive advantages. Increasing inequalities and environmental degradation will thus derive from the global extension of the reach of market logic. The same logic simultaneously encourages national specialization, reducing economic diversity and aggravating vulnerability. It also leads to increasing volumes of transports and concomitant emissions of greenhouse gases. The increasingly asymmetric flows of embodied labor, energy, materials, and land, which are obscured by market prices, become more unequal as the distance between producer and consumer grows.[2]

In the wake of the financial crisis of 2008, it has become increasingly acceptable to advocate more elaborate societal regulation of the economy. To counteract the runaway logic of unregulated private finance, there have been calls for innovation in *public* finance. A fundamental challenge for actors representing public interests is how to establish regulatory frameworks for alleviating risk and vulnerability without seriously discouraging private actors from innovative contributions that are of benefit

---

[2] As Ruggiero (2013) has shown, successive schools of economic thought from John Locke to neoliberalism have justified and encouraged various forms of social injustice. While this ideological function of economics may provoke our moral indignation, we need to remind ourselves that it is not so much a consequence of corruption and conspiracy as of the inertia of the money artifact to which we have delegated the operation of society.

for society at large. To progress in meeting this challenge, we need to emphasize some analytical distinctions that tend to be ignored in mainstream discourse, such as between market principles and capitalism, everyday local life versus global finance, and long-term sustainability and survival versus short-term gain. The societal objective must be to strike a balance between such distinct interests and concerns, which in my opinion means establishing ways of insulating them from each other, rather than allowing one to be absorbed by the other. We thus need to take seriously the recent surge of voices advocating degrowth, relocalization, postcapitalism, and the imperative of collectively changing the rules of the game.

A central question for those of us concerned with designing an economics for sustainability is whether the main concern of economic policy must be economic growth and maximum employment,[3] or if it might prioritize environmental and social resilience, justice, and long-term survival, even if this would mean less-than-maximum encouragement of growth and employment. How this question is answered hinges on our framing assumptions, for example about the nature of money, efficiency, and exchange. As recent global negotiations on climate change have focused on issues no less drastic than drafting policies for *saving the biosphere* from human destruction, we need to ask ourselves if there really are limits to how imaginative we can be in proposing policy recommendations for a sustainable economy.

The currently burgeoning discourse on degrowth has so far been modest and ambivalent about radically transforming the central institutions of modern economic life. There have been sporadic references to the possible benefits of basic income (Kallis 2011; Alexander 2015; Gerber 2015) and community currencies (Douthwaite 2012; Dittmer 2013, 2015), but no systematic attempt to outline how such reforms could be harnessed to the objectives of degrowth. With few exceptions (e.g., Douthwaite 2012), the prospects of alternative currency systems have not been favorably discussed by writers explicitly advocating degrowth. Although I share Dittmer's (2013, 2015) and Blanc's (2012) misgivings about the actual achievements of experiments with LETS and other forms of community currencies, I believe there is much to learn from these experiences that could help the degrowth movement envisage a

---

[3] Note that I here use *employment* in the conventional sense of formally salaried jobs. As my proposal will illustrate, there are alternative ways of mobilizing people to conduct useful and meaningful work.

concrete strategy for reaching its important objectives. Social experiments ranging from those of the utopian socialist Robert Owen in the early nineteenth century to the barter networks of Argentina in the beginning of the twenty-first century have all been founded on the insight that conventional money is destructive of both social and human-environmental relations (North 2007).

I cannot see any better strategy for economic *relocalization* – a frequently expressed goal among degrowth theorists – than to encode spatial distinctions into the currency system. To be able to imagine an economy that does not continue to generate increasing inequalities as well as greenhouse gas emissions, biodiversity loss, and other processes threatening to exceed planetary boundaries, we need to consider radically different ways of organizing human societies. This requires us to bracket conventional convictions about the requisites of a well-functioning economy, and to envisage feasible organizational changes that may modify human behavior patterns to increase sustainability and justice. We need nothing less than a *vision* of a more sustainable and egalitarian future human society, and a feasible road map for getting there without abandoning our values regarding democracy, justice, and individual liberty. If it is indeed easier to imagine the end of the world than the end of capitalism, it is high time to discover the elephant in the room.

Relocalization: A Proposal and Some Frequently Asked Questions

Given these challenges, we need to envisage how it would be possible to organize an economy that in at least three fundamental ways generates tendencies in directions diametrically opposite to those we observe in the world at present. Initial questions would be: How do we encourage human behavior patterns that *increase* rather than reduce sustainability, that *reduce* rather than increase vulnerability, and that *diminish* rather than increase inequalities? To accomplish the first of these goals, the objective must be to reduce transports, emissions, resource use, and waste. To promote the second goal, the ideal must be to enhance food security, diversity, community, and general resilience. To reach the third goal, finally, the aim must be to mitigate economic accumulation, polarization, and marginalization.

To encourage such modifications of human behavior without resorting to totalitarian politics and severe austerity measures, I propose the following general recommendation for a policy for sustainability and justice: *Each country establishes a complementary currency for local use only, which is distributed to all its residents as a basic income.* This general outline of a policy proposal raises a long list of questions to which only preliminary answers can be provided here. The specific ways in which these questions can be answered represent options for calibrating the proposal with different kinds of constraints, whether specific to different areas or of a more universal nature. The following attempt to provide preliminary responses to some of these questions should thus be understood as preliminary and subject to meticulous negotiation, monitoring, and modification:

1. *What is a complementary currency (CC)?* It is a form of money that can be used alongside regular currency, without in any way legally restricting the use of regular money.

2. *What are the fundamental goals of this proposal?* The two most fundamental goals motivating this proposal are to insulate local human subsistence and livelihood from the vicissitudes of national and international economic cycles and financial speculation, and to provide tangible and attractive incentives for people to live and consume more sustainably. It also seeks to provide authorities with a means to employ social security expenditures to channel consumption in sustainable directions and encourage economic diversity and community resilience at the local level. A system of separate currencies for basic needs versus global markets would insulate values pertaining to sustainability and survival from the abstract capital flows of the world-system. It could thus serve to check exploitation and environmental destruction. The central feature of the proposal is to restrain the spatial reach of the means of payment that propel substantial parts of a nation's social metabolism.

3. *Why should the state administrate the reform?* The nation is currently the most encompassing political entity capable of administrating an economic reform of this nature. Ideally, it is also subservient to the democratic decisions of its population. The current proposal is envisaged as an option for European nations but would seem equally advantageous for countries anywhere. If successfully implemented within a certain nation or set of nations, the system can be expected to be emulated by others. Whereas earlier experiments with alternative currencies have generally been local, bottom-up initiatives, a state-supported program offers advantages for long-term success. Rather than an informal,

marginal movement connected to identity politics within transient social networks, persisting only as long as the enthusiasm of its founders, the CC advocated here is formalized, efficacious, and lastingly fundamental to everyone's economy.

4. *How is local use defined and monitored?* The CC can only be used to purchase goods and services that are produced within a given geographical radius of the point of purchase. This radius can be defined in terms of kilometers of transport, and it can vary between different nations and regions depending on circumstances. A fairly simple way of distinguishing local from nonlocal commodities would be to label them according to transport distance, much as is currently done regarding, for instance, organic production methods or "fair trade." Such transport certification would, of course, imply different labeling in different locales. An efficient tool in measuring transport distances would be GPS (Global Positioning System) technology.

5. *How is the CC distributed?* A practical way of organizing distribution would be to provide each citizen with a plastic card that is electronically charged each month with the sum of CC allotted to him or her. The CC would thus be a completely electronic currency, and all transactions would be registered digitally.

6. *Who are included in the category of residents?* A monthly CC is provided to all inhabitants of a nation who have received official residence permits.

7. *What does basic income mean?* Basic income is distributed without any requirements or duties to be fulfilled by the recipients. The sum of CC paid to an individual each month can be determined in relation to the currency's purchasing power and to the individual's age. The guiding principle should be that the sum provided to each adult should be sufficient to enable basic existence, and that the sum provided for each child should correspond to the additional household expenses it represents.

8. *Why would people want to use their CC rather than regular money?* As the sum of CC provided each month would correspond to purchases representing a claim on his or her regular budget, the basic income would liberate a part of each person's regular income and thus amount to substantial purchasing power, albeit restricted only to local purchases. This incentive would at least initially – and paradoxically – be derived from opportunities for increased consumption, but the long-term, structural consequences of the

reform for the economy would entail a reduction in the consumption of commodities traded on the global market. However, the basic income in CC would reduce a person's dependence on wage labor and the risks currently associated with unemployment. It would encourage social cooperation and a vitalization of community.

9. *Why would businesses want to accept payment in CC?* Business entrepreneurs can be expected to respond rapidly to the radically expanded demand for local products and services, which would provide opportunities for a diverse range of local niche markets. Whether they receive all or only a part of their income in the form of CC, they can choose to use some of it to purchase local labor or other inputs, and to request to have some of it converted by the authorities to regular currency (see point 10).

10. *How is conversion of CC into regular currency organized?* All the recipients of CC would be granted the right to convert some of it into regular currency at exchange rates set by the authorities. The exchange rate between the two currencies should be calibrated to balance the in- and outflows of CC to the state. The rate would thus amount to a tool for determining the extent to which the CC is recirculated in the local economy or returned to the state. This is important to avoid inflation in the CC sector.

11. *Would there be interest on sums of CC owned or loaned?* There would be no interest accruing on a sum of CC, whether a surplus accumulating in an account or a loan extended.

12. *How would saving and loaning of CC be organized?* The formal granting of credit in CC would be managed by state authorities and follow the principle of full reserve banking, so that quantities of CC loaned would never exceed the quantities saved by the population as a whole.

13. *Would the circulation of CC be subjected to taxation?* Yes, but only when income in CC exceeds a certain threshold, indicating that the person in question is an entrepreneur engaged in accumulating CC beyond reasonable household consumption. This threshold, like the exchange rate referred to previously, can be continuously calibrated with other factors.

14. *Why would authorities want to encourage local economies that would deprive them of some of their tax income?* Given the beneficial social and ecological consequences of this reform, it is assumed that nation-states will represent the general interests of

their electorates and thus promote it. Particularly in a situation with rising fiscal deficits, unemployment, health care costs, and social security expenditures, the proposed reform would alleviate financial pressure on governments. It would also reduce the rising costs of transport infrastructure, environmental protection, carbon offsetting, and climate change adaptation. In short, the rising costs and diminishing returns on current strategies for economic growth can be expected to encourage politicians to consider proposals such as this, as a means of avoiding escalating debt or even bankruptcy.

15. *How would the state's expenditures in CC be financed?* The public costs of the reform would appear when companies and citizens request to convert some of their CC into regular money. As suggested previously, much of these expenditures would be balanced by the reduced costs for social security, health care, transport infrastructure, environmental protection, carbon offsetting, and climate change adaptation. As these savings may take time to materialize, however, states can choose to make a proportion of their social security payments – primarily pensions, sickness benefits, unemployment insurance, and family allowance – in the form of CC. This represents a significant option for financing the reform, requiring no corresponding tax levies. The proportions applicable to different kinds of social security payments should be determined in relation to how great a proportion of the recipient's total income is comprised by such payments, so that everyone has adequate access to both currencies.

16. *What are the differences between this CC and the many experiments with local currencies?* This proposal should not be confused with the idea and operation of local currencies, as it does not imply different currencies in different locales but *one* national CC for local use. Nor is it locally initiated and promoted in opposition to the regular currency, but centrally endorsed and administered as an accepted *complement* to it. Most importantly, the alternative currency can only be used to purchase products and services originating from within a given geographical range, a restriction that is not implemented in experiments with LETS. LETS have been rightly criticized for not challenging the formal market. Local currencies such as the Brixton pound do not restrain globalization; they can only be used in local shops, but in those shops, they can be converted into commodities from anywhere in the world.

In contrast to LETS and local currency schemes, this proposal introduces strong structural incentives to localize social metabolism, inhibiting globalization.[4] The basic income to which all should have a right should not be paid in general purpose money because that would simply keep us buying the products of the lowest-paid labor and the most degraded landscapes. It should be paid in a special purpose, complementary currency that can only be used to buy products and services produced within a specified radius of the point of purchase. This is not local money in the sense of geographically restricted currencies, but one single, national currency – issued by the state – that is only for local use. A final difference is that the CC is provided as a basic income to all residents of a nation, rather than only earned in proportion to the extent to which a person has made him- or herself useful in the local economy.

17. *What would the ecological benefits be?* The reform would radically reduce the demand for long-distance transports, the production of greenhouse gas emissions, consumption of energy and materials, and losses of foodstuffs through overproduction, storage, and transport. It would increase recycling of nutrients and packaging materials, which means decreasing leakage of nutrients and less garbage. It would reduce agricultural intensification, increase biodiversity, and decrease ecological degradation and vulnerability.

18. *What would the societal benefits be?* The reform would increase local cooperation and community cohesion, decrease social marginalization and addiction problems, provide more physical exercise, improve psycho-social and physical health, and increase food security and general community resilience. It would also decrease the number of traffic accidents, provide fresher and healthier food with less preservatives, and facilitate direct contact between producers and consumers. Given the recurrent indications that economic and cultural globalization also tends to disturb many people's sense of identity, a relocalization of social interaction

---

[4] In the same way that the practice of "earmarking" general purpose money for specific purposes (Zelizer 1998, 2017) fails to transform the fundamental inertia inherent in the generalized commensurability of such money, local currency schemes fail to counteract the inertia of globalization.

may engender a more secure experience of self, which in turn may alleviate ethnic and national tensions.

19. *What would the long-term consequences be for the economy?* The reform would no doubt generate radical transformations of the economy, as is precisely the intention. There would be a significant shift of dominance from transnational corporations founded on financial speculation and trade in industrially produced foodstuffs, fuels, and other internationally transported goods to locally diverse producers and services geared to sustainable livelihoods. This would be a democratic consequence of consumer power, rather than of legislation. Through a relatively simple transformation of the conditions for market rationality, governments can encourage new and more sustainable patterns of consumer behavior. In contrast to much of the drastic and often-traumatic economic change of the past two centuries, these changes would be democratic and sustainable, alleviate the exploitation of low-wage labor, reduce the asymmetry of global resource flows, and improve local and national resilience. It would provide those market actors in the Global North who do not want to encourage low-wage and environmentally degrading production processes in the South the possibility of refraining from exploiting such conditions, rather than inadvertently reinforcing the destructive consequences of globalization.

20. *Why should society want to encourage people to refrain from formal employment?* It is increasingly recognized that full or high employment should not be a goal in itself, particularly if it implies escalating environmental degradation and energy and material throughput. Well-founded calls are thus currently made for degrowth, that is, a reduction in the rate of production of goods and services that are conventionally quantified by economists as constitutive of gross domestic production. Whether formal unemployment is the result of financial decline, technological development, or intentional policy for sustainability, no modern nation can be expected to leave its citizens economically unsupported. To subsist on basic income is undoubtedly more edifying than receiving unemployment insurance; the CC system encourages useful community cooperation and creative activities rather than destructive behavior that may damage a person's health.

21. *Why should people receive an income without working?* As previously observed, modern nations will provide for their citizens

whether they are formally employed or not. The incentive to find employment should ideally not be propelled only by economic imperatives, but more by the desire to maintain a given identity and to contribute creatively to society. Personal liberty would be enhanced by a reform that makes it possible for people to choose to spend some of their time on creative activities that are not remunerated on the formal market, and to accept the tradeoff possibly implied by a somewhat lower economic standard. People can also be expected to devote a greater proportion of their time to community cooperation, earning additional CC, which means that they will contribute more to society – and experience less marginalization – than the currently unemployed.

22. *Would savings in CC be inheritable?* No.

23. *How would transport distances of products and services be controlled?* It is reasonable to expect the authorities to establish a special agency for monitoring and controlling transport distances. It seems unlikely that entrepreneurs would attempt to cheat the system by presenting distantly produced goods as locally produced, as we can expect income in regular currency – due to its limitless commensurability – generally to be preferable to income in CC. Such attempts would also entail transport costs that should make the cargo less competitive in relation to genuinely local produce, suggesting that the logic of local market mechanisms would by and large obviate the problem.

24. *How would differences in local conditions – such as climate, soils, and urbanism – be dealt with?* It is unavoidable that there would be significant variation between different locales in terms of the conditions for producing different kinds of goods. This means that relative local prices in CC for a given product can be expected to vary from place to place. This may in turn mean that consumption patterns will vary somewhat between locales, which is predictable and not necessarily a problem. A localization of resource flows can be expected to result in a more diverse pattern of calibration to local resource endowments, as in premodern contexts. The proposed system allows for considerable flexibility in terms of the geographical definition of what is categorized as local, depending on such conditions. In a fertile agricultural region, the radius for local produce may be defined, for instance, as 20 kilometers, whereas in a less fertile or urban area, it may be 50 kilometers. People living in urban centers are faced with a special challenge.

The reform would encourage an increased production of food-stuffs within and near urban areas, which in the long run should affect urban planning. Over time, more people may also choose to move to the countryside, where the range of subsistence goods that can be purchased with CC will tend to be greater. In the long run, the reform can be expected to encourage a better fit between the distribution of resources such as agricultural land and demography. This is fully in line with the intention of reducing long-distance transports of necessities.

25. *What would the consequences be if people converted resources from one currency sphere into products or services sold in another, or engaged in speculation between the two spheres?* It seems unfeasible to monitor and regulate the use of distant imports such as machinery and fuels in producing produce for local markets, but as production for local markets is remunerated in CC, this should constitute a disincentive to invest regular money in such production processes. Production for local consumption can thus be expected to rely mostly – and increasingly – on local labor and other resource inputs. As the option is always there to convert CC into regular money through the authorities, according to a variable exchange rate continuously calibrated by the nation's central bank, there would be no incentive to convert CC into general purpose money at a rate less favorable than that offered by the authorities. The only scenario that might encourage speculation would be if people converted ordinary money into CC at one point in time in the expectation that they would make a profit by reconverting the CC into regular money at a later point, when the exchange rate has become more favorable. Such speculation could be discouraged, for instance, by setting a limit to the amount of CC that can be converted by any given market actor.

26. *Is this a form of protectionism, that would provoke legal action in defense of free trade?* It would not be reasonable to classify this proposal as protectionism because it does not suggest taxing imports or in any other way creating legal obstacles to international trade. It merely offers people alternative modes of conversion and resource management, which do not compete on the same market as long-distance imports. Nor can it be viewed as centralized regulation, as it is entirely founded on the free choices of market actors. Should substantial numbers of people choose to utilize CC as a means of provisioning their households, it would

be an example of the "consumer power" celebrated by mainstream economists.

27. *Is the proposal realistic, considering the colossal power interests
that it challenges?* At the current historical moment, it is not at all
realistic, but there are many reasons to anticipate that conditions
will change. Within the course of the twenty-first century, the
world may experience social, ecological, and financial crises so
severe that leading politicians will have to devise ways of safe-
guarding their electorates that are radically alternative to relying
on economic growth and the global market. In such a situation,
I hope that some of them will be prepared to reconsider the kind of
ideas presented here. In that other light, this proposal will prove
more realistic than current illusions about carbon trading, negative emission technologies, geoengineering, overthrowing capitalist property relations, and other utopian schemes for surviving the
Anthropocene.

The biosphere will not be saved by trying to get the colonial outlook of
mainstream economics to somehow accommodate the Second Law of
Thermodynamics. Nor will it be saved by a socialist revolution that
collectivizes ownership but continues to think in terms of monetary value
and investments in advanced technologies. Money is not just a reflection
of the relations of production but an artifact that operates as an active
force in organizing such relations. To change the logic of the global
economic game, we must change the very artifact of money.

The ecological and societal destructiveness of what we call capitalism is
inherent in the cultural meme of general purpose money: the imperative of
minimizing costs and maximizing income. Given such money, the most
rational thing for market actors to do is to purchase the least expensive
commodities. In a local community context, the aggregate logic of such
behavior does seem to yield benefits for all concerned, as Adam Smith
envisaged. However, in a globalized economy with low transport costs
and considerable international discrepancies in the pricing of labor time
and other resources, market logic has tended to increase economic
inequalities and environmental degradation. Everyday consumption patterns have tended to favor the commodities embodying the lowest paid
labor, significant greenhouse gas emissions from transports, and the least
concern for the natural environment. If there are no constraints on what
we can buy with our money, we shall naturally be looking for the best
deals, which usually means the lowest-paid labor and the lowest-priced

resources. In choosing the least expensive products, consumers inadvertently encourage production processes that show the least concern for the environment, which often means the feeblest environmental legislation. The logic of the globalized capitalist market is an expression of the inherent properties of general purpose money. This logic unfolds along the same trajectories regardless of whether such money is created by banks or states, and even regardless of the existence of interest.

The use of conventional money at the global scale thus inexorably generates increasing inequalities, transports, greenhouse gas emissions, climate change, and environmental degradation. As these processes are consequences inherent in general purpose money, any attempt to curb the detrimental processes while continuing to comply with the rationality defined by such money will produce self-contradictory legislation raising the protests of economists and enterprise, such as protectionism and taxation aiming to inhibit economic activity. Given the conceptual constraints imposed by general purpose money and the resulting definition of market efficiency, mainstream economists should perhaps not be faulted for endorsing economic strategies that increase inequalities and threaten to make the planet uninhabitable for human beings. If they are to be criticized, it is because they have failed to envisage the feasibility and advantages of an alternative to general purpose money.

The predicament of the Anthropocene requires a redesigned currency system with a different logic, providing market actors with a special purpose money that incites them to purchase commodities embodying local labor, a minimum of transports and greenhouse gas emissions, and concern for the local environment. Such a voluntary localization of social metabolism would represent a democratic transformation of the structural conditions for market rationality. Given the new, localizing rationality introduced by the option of using a complementary currency as outlined here, the notion of consumer power would finally approximate the benevolent aspirations of market doctrine professed since the eighteenth century, generating more equitable and sustainable practices as a logical consequence of how the currency system is designed.

Afterword

Confronting Mainstream Notions of Progress

> Ideas are not powerful because they are true, Foucault insists, they are true
> because of power.
>
> Paul Robbins (2012: 124)

> It is the responsibility of intellectuals to speak the truth and to expose lies.
>
> Noam Chomsky (1967)

Is the world heading for disaster, or getting better and better? Considering
the many reasons we have to worry about the future of humanity and the
biosphere (cf. Latouche 2009),[1] a major puzzle for me is how so many
people seriously seem to believe that we can trust that technology and the
market will solve problems of global justice and ecological sustainability.
Out of curiosity, I have thus followed the late Hans Rosling's recommendation in *Factfulness* (2018) and seriously engaged with perspectives
largely opposite to my own. After decades of efforts to understand why
the world economy is generating abysmal inequalities, environmental
degradation, and climate change, I have thus carefully considered the
convictions of the so-called New Optimists. Besides Rosling, the most
prominent writer in this genre is Steven Pinker, whose book *Enlightenment Now: The Case for Reason, Science, Humanism and Progress*
(2018) argues that the world is getting better and better. This is also the

---

[1] Readers will already be familiar with some of the many alarming ecological and socio-
economic problems confronting modern civilization, many of which have been mentioned
in previous chapters, but in introducing his book on the necessity of degrowth, Latouche
(2009: 2) offers an appropriate conclusion: "Where are we going? We are heading for a
crash. We are in a performance car that has no driver, no reverse gear and no brakes and it
is going to slam into the limitations of the planet."

message of Johan Norberg's book *Progress: Ten Reasons to Look Forward to the Future* (2016). The three authors pursue similar arguments and frequently refer to each other.

I admit initially feeling suspicious about books strongly endorsed by Bill Gates. The formerly wealthiest man in the world describes *Factfulness* as "one of the most important books I've ever read" and *Enlightenment Now* as "my new favourite book of all time." It could hardly be a coincidence that books celebrating the current trends of world society are praised by one of its foremost beneficiaries. But it would be unfair to reduce Rosling's and Pinker's observations to mere ideology. So, let us start with the things we agree on, and then focus on what the New Optimists tend to disregard.

Rosling, Pinker, and Norberg all emphasize that, by and large, people all over the world throughout the past two centuries have experienced improving health, security, education, and living standards, while the global population has almost quadrupled. This is an incontrovertible fact. Pinker makes a strong case for science and rationality as the source of these improvements. In this respect, too, I completely agree. As I have argued in Chapters 10, 11, and 12, the currently fashionable "posthumanism" in much of the humanities and social sciences indeed frequently represents a descent into mysticism and muddled thinking.

But there are some huge lacunae in the worldview of the New Optimists. Rosling's observations on the improving medical condition of the world's population are presented as the only facts we need. He advocates a pragmatic mix of market and state but appears to consider a macroscale grasp of the world economy superfluous to his analysis. Although he urges us to think about how the "system" operates, the complete absence of social science is disturbing. This is as conspicuous in his account of historical trends as in his treatment of contemporary inequalities. Most of his historical graphs begin in the year 1800, after centuries of a capitalist world-system based on European exploitation, slavery, and violent conquest. But there is no mention of colonialism. To measure progress in relation to a point in time that for most of the world was an extreme low-water mark is to make the misleading assumption that average quality of life in 1800 was representative of the human condition up until then. If average Africans had *not* improved their lot somewhat since the heyday of colonialism, world inequality would have been even more scandalous – and politically volatile – than it is today. But the New Optimists celebrate an extremely unequal world economy by demonstrating that inequalities are not quite as obscene as they were 200 years ago. Moreover, as

Rosling's recurrent advice to corporations makes clear, a somewhat greater purchasing power, health, and education in the periphery is obviously in the interests of a global capitalism in pursuit of new markets and robust labor. But is this an indication of the benevolence of neoliberalism?

Even more problematic is the account of global inequalities today. Seven billion people in 2017 are represented as lined up in a queue progressing through four Rostowian stages defined in terms of income and living standard, from extreme poverty (less than US$2 per person and day) to affluence (more than US$32 per person and day). One billion people do not even have shoes, three billion can at most buy a bicycle, two billion have managed to invest in running water, electricity, and a refrigerator (and they can buy a motorcycle!) while only the top billion people are able to buy a car. In Rosling's narrative – as in Pinker's and Norberg's – people are steadily progressing toward the topmost level, and progress is just a matter of "catching up," as if global inequalities were differences in time rather than in space. There is no actual gap between "rich" and "poor" – apparently a figment of distorted binary thinking – but a continuum of seven billion people all heading in the same direction. This account would seem to compel us to continue doing whatever we have been doing for the past 200 years. But what Rosling misses is that those seven puppets animating his diagrams and presentations, each representing one billion people, don't just stand in line but have societal relations to each other. They are parts of the same global system, and as we have argued in previous chapters, there *are* contradictions and power relations in that system. The low wages of levels 2 and 3 are what make it possible for people at level 4 to buy a car – and why there are no cars on levels 2 and 3. Even if there is no neat dualism, there certainly is a polarization of rich and poor, and it is promoted by globalization. As countries like China have realized, low wages are good for business. But again, there is no mention of asymmetries in international trade. For those of us who can afford to buy a used car, the six billion majority who *cannot* do so will continue to strike us as "poor." Redefining them as "middle income" will not make global class polarities go away. According to Rosling's own statistics, a four billion majority of the world's population does not even have running water, sewerage, or waterproof roofing. The incontrovertible improvements in health and education have not dissolved inequalities. Contrary to the assertions of New Optimists like Rosling, Pinker, and Norberg, *The World Inequality Report 2018* shows that, between 1980 and 2016, global inequalities have increased (Alvaredo et al. 2017). It is part of the logic of the globalized world market, rather than an evil from which the market will save us.

No, statistical "facts" do not suffice to explain the world. For one thing, the way they are organized can distort and obscure the underlying relations. The four income levels that organize Rosling's data are arbitrary categories. Why draw a line between people earning US\$31 and US\$32 per day, but include in "level 4" the extremely diverse range of people from those who can barely afford to buy a used car to multibillionaires like Bill Gates? Such categories do not help us understand how Bill Gates has made his fortune through the employment of masses of low-wage labor throughout the world.

Confronting widespread worries and grievances about the development of world society and the biosphere, the New Optimists are unabashed apologetics for business as usual. When Björn Lomborg in *The Skeptical Environmentalist* (2001) organized statistics to argue that *all* the major environmental problems of the modern world were less problematic than generally believed, many of us sensed that he was masquerading a politically biased view as scientific research. Similarly, the New Optimists do not hesitate to let their uniformly reassuring messages cover the whole range of parameters from health, peace, education, and living standards to population growth, energy, biodiversity, and climate change. Such unreasonable claims to know just about everything better than the much-maligned skeptics result in startling arguments like Rosling's triumphant observation that tigers, pandas, and rhinos are not closer to extinction in 2017 than they were in 1996. Should this make us less worried about what Elizabeth Kolbert refers to as *The Sixth Extinction* (2014)? Or about the nine "planetary boundaries" to which Johan Rockström and colleagues (Rockström et al. 2009) have alerted us? We must be prepared to grasp that the unevenly distributed human improvements in health, security, education, and technology have come at a cost in terms of the future viability of the biosphere. The rising collective affluence of our species in 2017 should not be confused with the likelihood of its survival two centuries from now. Which of these issues has higher priority for the New Optimists? To celebrate that some things have been improving for a couple of centuries does not give us reason to assume that quite different things will improve in the future. Human history has no more of a transcendent direction than biological evolution. The question if the world is getting better thus cannot be answered with a simple yes or no.

Looking at a satellite image of nighttime lights is like looking at the kind of figure-ground drawings used to illustrate optical illusions. In one such drawing, you can either see a rabbit or a bird. In the satellite images, you can see economic and technological growth either as a cornucopia or as the accumulation of resources through appropriation from other parts of the world. Because successful environmental policies correlate with growth, it is easy to assume that growth contributes to environmental quality. Rosling, Pinker, Norberg, and other neoliberal optimists indeed think of economic growth and technological intensification as good for the environment, rather than a force of destruction. The higher the GDP, the better for the environment. But in this book we have argued that it is the other way around: the engines of human progress are simultaneously the source of our growing vulnerability. A second look at the satellite image might help us see another pattern, in which the world market and globalized technologies make it possible for richer nations to displace work and environmental pressure to poorer ones. Rather than blame the poor for the deterioration of the biosphere, we might realize that they are the ones who do the dirty work of producing cheap commodities for us to consume. The "facts" are the same, but the interpretation is diametrically different. Pinker (2018: 403) warns that we should never infer causation from a correlation, but like other neoliberal ecomodernists he views technological intensification – epitomized by the luminous, urban areas in the satellite images – as the key to global justice and sustainability. One of his rosy convictions that can be decisively dismissed is that technological progress is propelling "dematerialization" – that we can now "do more with less" – an assertion that is convincingly contradicted by the recent report from the UN Environmental Programme (Schandl et al. 2016), which shows that the material intensity of economic growth has in fact *increased* over the past decade.

When it comes to mitigating global warming, it is hardly surprising that Pinker's (2018: 146) main solution is nuclear power. The almost universal trust in technological progress – most dramatically illustrated by Pinker's and Norberg's (and Bill Gates's) imaginative visions of perpetual-motion reactors fueled by their own waste (ibid., 149) – is based on an incomplete grasp of what modern technologies really *are*, from a global perspective. The standard story is that smart people since the Industrial Revolution have discovered new ways of harnessing energy in nature. But this does not include the inequalities of the world economy as part of the conditions for developing the means to harness such energy. As we have seen, the expansion of steam power in eighteenth- and nineteenth-century Britain would have been impossible without the Atlantic slave trade, the colonial cotton plantations, and the accumulation of capital in the core of the British Empire. Workloads and environmental loads were largely displaced to the periphery, whose inexpensive labor and land were used to produce commodities that brought wealth to merchants and factory owners. As parts of this wealth was invested in new machinery, global flows of embodied labor and land further aggravated the asymmetric relation between core and periphery. To this day, you must have lots of money to develop and adopt new technology. It is not a coincidence that shifts to renewable energy have progressed furthest in the wealthier countries of the world. This means that technologies and energy use are not politically neutral, "natural" phenomena but have a societal, distributive aspect that both economists and engineers tend to overlook. As has become obvious in recent years, the European adoption of photovoltaic panels or cars propelled by sugarcane ethanol hinges on the prices of Chinese labor and Brazilian land. This is not to deny that technological progress has brought immeasurable human benefits since the eighteenth century, but to maintain – contrary to the New Optimists and ecomodernists – that it has continuously imposed unacknowledged trade-offs in terms of social inequalities and ecological degradation. Modern technologies tend to save time and space for those who can afford them, but – and this applies even to Rosling's celebrated washing machine – at the expense of human time and natural space lost for those who cannot. Again, this is not to advocate Luddite machine-breaking, but to expand our understanding of technology as a total socioecological phenomenon that is as contingent on relative market prices as on human ingenuity. It is not to descend into obscurantism, but to turn Enlightenment against itself by identifying the limits of modern rationality. It is certainly not rational or enlightened to allow the logic of mainstream economics to continue to encourage the dismantling of the biosphere. As Pinker (ibid., 378) observes, "[T]he beauty of reason is that it can always be applied to understand failures of reason." I am ultimately optimistic in the sense that I trust that currently mainstream conceptions of rationality – which means delegating the future of the world to the fetishism of the market – will be reassessed as profoundly irrational. *This* is the Enlightenment we now need.

There is another source of optimism that can be gleaned from the figures and graphs presented by the New Optimists. Rosling's congratulations to the five billion people who populate levels 2 and 3 confirm that healthy and meaningful lives can be lived with a small fraction of the income available to those on level 4. Considering the vast ecological footprints and huge carbon emissions associated with the latter's exuberant lifestyles, this is nothing less than an argument for *degrowth*. The wealthiest billion people are as much of a problem as the billion without shoes. But this is hardly the message intended by growth enthusiasts like Pinker and Norberg.

Pinker's excursions into political theory are obviously tainted by his own neoliberal worldview. Although recent decades of globalization have generated at least three distinct political ideologies – liberals endorsing the market, leftists demanding justice, and populists guarding their spatially anchored, national identities – Pinker (2018: 333–334) explicitly conflates leftist critiques of capital and xenophobic attacks on immigrants as different brands of "authoritarian populism." This is a noteworthy mirror image of the leftists' inclination to lump neoliberals and populists as "bourgeois" and the populists' propensity to lump leftists and liberals as "globalists." A triad is everywhere reduced to a dualism. But socialists are no doubt as disturbed as liberals to find themselves lumped with xenophobes, while most nationalists would resent being classed with either leftists or liberals. The quintessentially socialist concern with justice is difficult to fit into the increasingly pronounced polarization of globalists versus nationalists. Socialists have long challenged globalization as inherently conducive to injustice, while nationalism – although an integral ingredient of all the socialist projects of the twentieth century – now suggests right-wing sentiments and the justification of domestic as well as global class differences. As the political landscape is being reorganized, we may conclude, it has become increasingly difficult to challenge the inequalities that are inherent in the neoliberal world market and advanced technologies. The ideological hegemony of globalization and ecomodernism is as evident in the mainstream rejection of dissidents as "technophobes" as in the risk they run of being rejected as xenophobes. The neoliberal reconstitution of capitalist ideology appears to have effectively immunized the globalized market against objections of any kind, whether skepticism is directed at economic growth, purportedly "green" technologies contingent on the exploitation of global price differences, or the disavowal of cultural identity. Unwittingly, in their endorsement of advanced technologies (e.g., Schwartzman 1996) and their rejection of place-based identities, even socialists are now promoting neoliberalism. Although ostensibly radical critics of the "system," even the anticapitalists are ultimately reproducing it.

In the academic sphere, too, a triad of prominent positions has complicated debate on global society. Each such position deserves to be assessed separately because they all demonstrate insights as well as blind spots. The liberal ecomodernists represented by Steven Pinker champion reason and objectivism but devote very little attention to justice, exploitation, and global power structures. Marxists like David Harvey emphasize justice, exploitation, and power but tend to converge with the ecomodernists in subscribing to a Promethean understanding of technology based on the illusion that I have called *machine fetishism*. Finally, posthumanists like Bruno Latour deconstruct technologies and other artifacts as reifications of social relations rather than morally neutral gifts of nature but tend ultimately to endorse fetishism and show conspicuously little interest in justice, exploitation, and power. Given their common dedication to reason and science, the liberal ecomodernists and the Marxists in this respect are opposed to the posthumanists. In contrast to Marxists, however, ecomodernists and posthumanists are equally unconcerned with issues of justice and power. However, Marxists and posthumanists both challenge the liberals by being critical of what the latter celebrate as progress. The position that I have adopted in this book seeks to integrate the premises of these different approaches by combining a rational critique of power with a fundamental, anthropological readiness to deconstruct and defamiliarize our everyday categories.

It is paradoxical to find neoliberal optimists like Pinker and Norberg lamenting the recent expansion of populism and protectionist trade policies. The future suddenly seems less bright than they have anticipated. Moreover, the very concept of "liberal" needs unpacking. For Pinker, to be "liberal" in the moral sense of advocating democracy, tolerance, and human rights seems automatically synonymous with being "liberal" in the sense of endorsing free trade, but moral codes and economic policies cannot be so simply equated. Considering what globalization has been doing to peripheral sectors of world society since the nineteenth century, it is highly misleading to claim that core humanist values such as the Golden Rule[2] are connected to liberal trade policies. The history of Europe's role in the world hardly qualifies it as the bastion of human rights, as many European politicians would have it. Equally misleading is Pinker's (2018: 396–397) conflation of Marxist critical theory and postmodernist social science. Most Marxists would no doubt share Pinker's endorsement of Enlightenment concepts of truth and rationality. But the

---

[2] The Golden Rule is a general injunction of reciprocity that occurs in many religions and cultural contexts, but its common essence is that one should treat others as one would like others to treat oneself.

Enlightenment that we must now hope for will pivot on our ability to move beyond the notion of science as a purified understanding of the physical world as if it were distinct from politics and inequalities. In this book I have argued that we need to develop our understanding of how natural phenomena and social organization are intertwined – most importantly, fossil energy and the money fetishism we know as capitalism. The illusion promoted by ideologists from Adam Smith to the New Optimists is that market exchange and engineering are politically neutral phenomena that do *not* add up to a mode of exploitation, in other words that what we call capitalism does not violate core humanist values. This conviction, of course, is utterly political.

References

Alexander, Samuel. 2015. Basic and maximum income. In *Degrowth: A vocabulary for a new era*, edited by Giacomo D'Alisa, Federico Demaria, and Giorgos Kallis, 146–148. London: Routledge.

Altvater, Elmar. 1994. Ecological and economic modalities of time and space. In *Is capitalism sustainable? Political economy and the politics of ecology*, edited by Martin O'Connor, 76–90. New York: Guilford.

— 2007. The social and natural environment of fossil capitalism. *Socialist Register* 43: 37–59.

— 2016. The Capitalocene, or, geoengineering against capitalism's planetary boundaries. In *Anthropocene or Capitalocene? Nature, history, and the crisis of capitalism*, edited by Jason W. Moore, 138–152. Oakland, CA: PM Press.

Alvaredo, Facundo, Lucas Chancel, Thomas Piketty, Emmanuel Saez, and Gabriel Zucman. 2017. *World inequality report 2018*. Paris: World Inequality Lab.

Amin, Samir. 1976. *Unequal development*. New York: Monthly Review Press.

Andersen, Otto. 2013. *Unintended consequences of renewable energy*. London: Springer.

Angus, Ian. 2016. *Facing the Anthropocene: Fossil capitalism and the crisis of the Earth System*. New York: Monthly Review Press.

Aristotle. 1962. *The politics*. Harmondsworth, UK: Penguin.

Asafu-Adjaye, John, Linus Blomqvist, Stewart Brand, Barry Brook, Ruth DeFries, Erle Ellis, Christopher Foreman et al. 2015. *An ecomodernist manifesto*. www.ecomodernism.org.

Barca, Stefania. 2011. Energy, property, and the Industrial Revolution narrative. *Ecological Economics* 70: 1309–1315.

Barth, Fredrik, ed. 1998. *Ethnic groups and boundaries: The social organization of culture difference*. Long Grove, IL: Waveland Press.

Bateson, Gregory. 1972. *Steps to an ecology of mind*. Frogmore, UK: Paladin.

Baudrillard, Jean. [1972] 1981. *For a critique of the political economy of the sign.* St. Louis, MO: Telos.

Bauman, Zygmunt. 1998. *Globalization: The human consequences.* Cambridge: Polity.

2011. *Collateral damage: Social inequalities in a global age.* Cambridge: Polity.

Beck, Ulrich. 1992. *Risk society: Towards a new modernity.* London: Sage.

Beckert, Sven. 2014. *Empire of cotton: A global history.* New York: Vintage Books.

Bennett, Jane. 2010. *Vibrant matter: A political ecology of things.* Durham, NC: Duke University Press.

Benton, Ted. 1989. Marxism and natural limits: An ecological critique and reconstruction. *New Left Review* 178: 51–86.

Berg, Maxine. 1980. *The machinery question and the making of political economy 1815–1848.* Cambridge: Cambridge University Press.

Besley, Tim, and Peter Hennessy. 2009. The global financial crisis – why didn't anybody notice? *British Academy Review* 14: 8–10.

Bessire, Lucas, and David Bond. 2014. Ontological anthropology and the deferral of critique. *American Ethnologist* 41(3): 440–456.

Bhaskar, Roy. 1975. *A realist theory of science.* Leeds, UK: Leeds Books.

Bijker, Wiebe E. 2009. Globalization and vulnerability: Challenges and opportunities for SHOT around its fiftieth anniversary. *Technology and Culture* 50(3): 600–612.

2010. How is technology made? That is the question! *Cambridge Journal of Economics* 34(1): 63–76.

Bijker, Wiebe E., Thomas P. Hughes, and Trevor J. Pinch, eds. 1987. *The social construction of technological systems: New directions in the sociology and history of technology.* Cambridge, MA: MIT Press.

Bildt, Carl. 2016. It's the end of the West as we know it. *Washington Post,* November 15 .

Bjerg, Ole. 2014. *Making money: The philosophy of crisis capitalism.* London: Verso.

Blanc, Jérôme. 2012. Thirty years of community and complementary currencies: A review of impacts, potential and challenges. *International Journal of Community Currency Research* 16: 1–4.

Blaser, Mario. 2013. Ontological conflicts and the stories of peoples in spite of Europe: Toward a conversation on political ontology. *Current Anthropology* 54(5): 547–568.

Blaut, James M. 2000. *Eight Eurocentric historians.* New York: Guilford Press.

Bloch, Maurice, and Jonathan Parry. 1989. Introduction: Money and the morality of exchange. In *Money and the morality of exchange,* edited by Jonathan Parry and Maurice Bloch, 1–32. Cambridge: Cambridge University Press.

Bohannan, Paul. 1955. Some principles of exchange and investment among the Tiv. *American Anthropologist* 57: 60–70.

Bonneuil, Christophe, and Jean-Baptiste Fressoz. 2015. *The shock of the Anthropocene: The Earth, history and us.* London: Verso.

Bourdieu, Pierre. 1991. *Language and symbolic power*. Translated by Gino Raymond and Matthew Adamson. Cambridge, MA: Harvard University Press.

Braudel, Fernand. 1992. *The perspective of the world: Civilization and capitalism, 15th–18th centuries*, vol. 3. Berkeley: University of California Press.

Brennan, Teresa. 1997. Economy for the Earth: The labour theory of value without the subject/object distinction. *Ecological Economics* 20: 175–185.

Bridge, Gavin, Stefan Bouzarovski, Michael Bradshaw, and Nick Eyre. 2013. Geographies of energy transition: Space, place and the low-carbon economy. *Energy Policy* 53: 331–340.

Broome, John. 2015. Do not ask for morality. Paper presented at the conference How to Think the Anthropocene, Paris, November 2015.

Bryant, Raymond L., ed. 2015. *The international handbook of political ecology*. Cheltenham, UK: Edward Elgar.

Bryant, Raymond L., and Sinead Bailey. 1997. *Third World political ecology*. London: Routledge.

Buber, Martin. 1970. *I and Thou*. New York: Scribner's.

Buck, Holly Jean. 2015. On the possibilities of a charming Anthropocene. *Annals of the Association of American Geographers* 105(2): 369–377.

Bunker, Stephen G. 1985. *Underdeveloping the Amazon: Extraction, unequal exchange, and the failure of the modern state*. Chicago: University of Chicago Press.

Burkett, Paul. 1999. Fusing red and green. *Monthly Review* 50 (9): 47–56.

——— [2005] 2009. *Marxism and ecological economics: Toward a red and green political economy*. Leiden, The Netherlands: Brill.

——— [1999] 2014. *Marx and nature: A red and green perspective*. Chicago: Haymarket.

Burkett, Paul, and John B. Foster. 2006. Metabolism, energy, and entropy in Marx's critique of political economy: Beyond the Podolinsky myth. *Theory and Society* 35: 109–156.

Büscher, Bram, Robert Fletcher, Dan Brockington, Chris Sandbrook, William M. Adams, Lisa Campbell, Catherine Corson, Wolfram Dressler, Rosaleen Duffy, Noella Gray, George Holmes, Alice Kelly, Elisabeth Lunstrum, Maano Ramutsindela, and Kartik Shanker. 2016. Half-Earth or whole Earth? Radical ideas for conservation, and their implications. *Oryx* 51(3): 407–410.

Callon, Michel. 1987. Society in the making: The study of technology as a tool for sociological analysis. In *The social construction of technological systems: New directions in the sociology and history of technology*, edited by Wiebe E. Bijker, Thomas P. Hughes, and Trevor J. Pinch, 77–97. Cambridge, MA: MIT Press.

——— ed. 1998. *The laws of the markets*. Oxford: Blackwell Publishers.

Carson, Rachel. 1962. *Silent spring*. Boston: Houghton Mifflin.

Castree, Noel. 2014. *Making sense of nature: Representation, politics and democracy*. London: Routledge.

Chakrabarty, Dipesh. 2009. The climate of history: Four theses. *Critical Inquiry* 35: 197–222.

2014. Video presentation at the conference *Os mil nomes de Gaia*, Rio de Janeiro, September 15–19.

2015. The human significance of the Anthropocene. *Manuscript.*

Chase-Dunn, Christopher, and Thomas D. Hall. 1997. *Rise and demise: Comparing world-systems.* Boulder, CO: Westview.

Chomsky, Noam. 1967. The responsibility of intellectuals. *New York Review of Books*, February 23.

Costanza, Robert. 1980. Embodied energy and economic evaluation. *Science* 210: 1219–1224.

Costanza, Robert, Ralph d'Arge, Rudolf de Groot, Stephen Farber, Monica Grasso, Bruce Hannon, Karin Limburg, Shahid Naeem, Robert V. O'Neill, Jose Paruelo, Robert G. Raskin, Paul Sutton, and Marjan van den Belt. 1997. The value of the world's ecosystem services and natural capital. *Nature* 387: 253–260.

Crist, Eileen. 2016. On the poverty of our nomenclature. In *Anthropocene or Capitalocene? Nature, history, and the crisis of capitalism*, edited by Jason W. Moore, 14–33. Oakland, CA: PM Press.

Cronon, William. 1995. The trouble with wilderness; or, getting back to the wrong nature. In *Uncommon ground: Rethinking the human place in nature*, edited by William Cronon, 69–90. New York: W. W. Norton.

Crutzen, Paul J., and Eugene F. Stoermer. 2000. The "Anthropocene." *IGBP Newsletter* 41: 17–18.

D'Alisa, Giacomo, Federico Demaria, and Giorgos Kallis. 2014. *Degrowth: A vocabulary for a new era.* London: Routledge.

Daly, Herman E. 1996. *Beyond growth.* Boston: Beacon Press.

Daly, Herman E. 2018. Ecologies of scale: Interview by Benjamin Kunkel. *New Left Review* 109: 81–104.

Danowski, Déborah, and Eduardo Viveiros de Castro. 2017. *The ends of the world.* Cambridge: Polity.

Davies, Glyn. 2002. *A history of money.* Cardiff: University of Wales Press.

Davies, Jeremy. 2016. *The birth of the Anthropocene.* Oakland: University of California Press.

Debeir, Jean-Claude, Jean-Paul Deléage, and Daniel Hémery. 1991. *In the servitude of power: Energy and civilization through the ages.* London: Zed Books.

de Brunhoff, Suzanne. [1973] 2015. *Marx on money.* London: Verso.

de la Cadena, Marisol. 2010. Indigenous cosmopolitics in the Andes: Conceptual reflections beyond "politics." *Cultural Anthropology* 25(2): 334–370.

2015. *Earth beings: Ecologies of practice across Andean worlds.* Durham, NC: Duke University Press.

Deléage, Jean-Paul. 1989. Eco-Marxist critique of political economy. *Capitalism Nature Socialism* 1 (3): 15–31.

Descola, Philippe. 2013. *Beyond nature and culture.* Chicago: University of Chicago Press.

Descola, Philippe, and Gísli Pálsson, eds. 1996. *Nature and society: Anthropological perspectives.* London: Routledge.

Dittmer, Kristofer. 2013. Local currencies for purposive degrowth? A quality check of some proposals for changing money-as-usual. *Journal of Cleaner Production* 54: 3–13.

2015. Community currencies. In *Degrowth: A Vocabulary for a new era*, edited by Giacomo D'Alisa, Federico Demaria, and Giorgos Kallis, 149–151. London: Routledge.

Dorninger, Christian, and Alf Hornborg. 2015. Can EEMRIO analyses establish the occurrence of ecologically unequal exchange? *Ecological Economics* 119: 414–418.

Douthwaite, Richard. 2012. Degrowth and the supply of money in an energy-scarce world. *Ecological Economics* 84: 187–193.

Elder-Vass, Dave. 2008. Searching for realism, structure and agency in Actor Network Theory. *The British Journal of Sociology* 59(3): 455–473.

2015. Disassembling actor-network theory. *Philosophy of the Social Sciences* 45(1): 100–121.

Ellen, Roy F. 1996. The cognitive geometry of nature: A contextual approach. In *Nature and society: Anthropological perspectives*, edited by Philippe Descola and Gísli Pálsson, 103–123. London: Routledge.

Ellul, Jacques. 1964. *The technological society*. New York: Knopf.

Emmanuel, Arghiri. 1972. *Unequal exchange: A study of the imperialism of trade*. New York: Monthly Review Press.

Ferguson, Niall. 2008. *The ascent of money: A financial history of the world*. New York: Penguin.

Foster, John B. 2000. *Marx's ecology: Materialism and nature*. New York: Monthly Review Press.

Foster, John B. 2014. Foreword. In Paul Burkett, *Marx and nature: A red and green perspective*, vi–xiii. Chicago: Haymarket.

Foster, John B., and Paul Burkett. 2004. Ecological economics and classical Marxism: The "Podolinsky business" reconsidered. *Organization and Environment* 17(1): 32–60.

2008. Classical Marxism and the Second Law of Thermodynamics: Marx/Engels, the heat death of the universe hypothesis, and the origins of ecological economics. *Organization and Environment* 21(1): 3–37.

2016. *Marx and the Earth: An anti-critique*. Leiden, The Netherlands: Brill.

Foster, John B., and Brett Clark. 2016. Marx's ecology and the Left. *Monthly Review* 68(2): 1–25.

Foster, John B., Brett Clark, and Richard York. 2010. *The ecological rift: Capitalism's war on the Earth*. New York: Monthly Review Press.

Foster, John B., and Hanna Holleman. 2014. The theory of unequal ecological exchange: A Marx-Odum dialectic. *The Journal of Peasant Studies* 41(2): 199–233.

Frank, Andre G. 1998. *ReOrient: Global economy in the Asian age*. Berkeley: University of California Press.

Frank, Andre G., and Barry K. Gills, eds. 1993. *The world system: Five hundred years or five thousand?* London: Routledge.

Frey, R. Scott, Paul K. Gellert, and Harry F. Dahms, eds. 2017. Special issue: Unequal ecological exchange. *Journal of World-Systems Research* 23(2): 226–398.

eds. 2019. *Ecologically unequal exchange: Environmental injustice in comparative and historical perspective*. Houndmills, UK: Palgrave Macmillan.

Friedman, Jonathan. 1974. Marxism, structuralism and vulgar materialism. Man *(New Series)* 3: 444–469.

Gellert, Paul K. 2019. Bunker's ecologically unequal exchange, Foster's metabolic rift, and Moore's world-ecology: Distinctions with or without a difference? In *Ecologically unequal exchange: Environmental injustice in comparative and historical perspective*, edited by Scott R. Frey, Paul K. Gellert, and Harry F. Dahms, 107–140. Houndmills, UK: Palgrave Macmillan.

Georgescu-Roegen, Nicholas. 1971. *The Entropy Law and the economic process.* Cambridge, MA: Harvard University Press.

1979. Energy analysis and economic valuation. *Southern Economic Journal* 45: 1023–1058.

1986. The entropy law and the economic process in retrospect. *Eastern Economic Journal* XII: 3–25.

Gerber, Julien-Francois. 2015. An overview of local credit systems and their implications for post-growth. *Sustainability Science* 10: 413–423.

Godelier, Maurice. 1969. La monnaie de sel des Baruya de Nouvelle-Guinée. *L'Homme* IX(2): 5–37.

1978. Infrastructures, societies and history. *Current Anthropology* 19: 763–771.

1994. "Mirror, mirror on the wall..." The once and future role of anthropology: A tentative assessment. In *Assessing cultural anthropology*, edited by Robert Borofsky, 97–109. New York: McGraw-Hill.

Goldman, Mario. 2009. An Afro-Brazilian theory of the creative process: An essay in anthropological symmetrization. *Social Analysis* 53(2): 108–129.

Goodhart, David. 2017. *The road to somewhere: The new tribes shaping British politics.* London: Penguin Books.

Gorz, André. 1994. *Capitalism, socialism, ecology.* London: Verso.

Graeber, David. 2011. Debt: The first 5,000 years. Brooklyn, NY: Melville House.

Gregory, Chris A. 1982. *Gifts and commodities.* London: Academic Press.

2014. On religiosity and commercial life: Toward a critique of cultural economy and posthumanist value theory. *HAU: Journal of Ethnographic Theory* 4(3): 45–68.

Gren, Martin. 2009. Time geography. In *The international encyclopedia of human geography*, edited by Robert Kitchin and Nigel Thrift, 279–284. Amsterdam, The Netherlands: Elsevier.

Hagens, Nathan John. 2015. Energy, credit, and the end of growth. In *State of the world 2015: Confronting hidden threats to sustainability*, 21–35. The Worldwatch Institute. Washington, DC: Island Press.

Hall, Charles A. S., and Kent A. Klitgaard. 2011. *Energy and the wealth of nations: Understanding the biophysical economy.* New York: Springer.

Hamilton, Clive. 2003. *Growth fetish.* London: Pluto Press.

2010. *Requiem for a species: Why we resist the truth about climate change.* London: Earthscan.

2015. Towards a philosophy of history for the Anthropocene. Paper presented at the conference Comment penser l'anthropocene? Collège de France, Paris, November 5–6.

2017. *Defiant Earth: The fate of humans in the Anthropocene.* Cambridge: Polity.

Haraway, Donna. 1988. Situated knowledges: The science question in feminism and the privilege of partial perspective. *Feminist Studies* 14(3): 575–599.

2007. *When species meet*. Minneapolis: University of Minnesota Press.

2015. Anthropocene, Capitalocene, Plantationocene, Chthulucene: Making kin. *Environmental Humanities* 6: 159–165.

2016. *Staying with the trouble: Making kin in the Chthulucene*. Durham, NC: Duke University Press.

Harman, Graham. 2007. The importance of Bruno Latour for philosophy. *Cultural Studies Review* 13(1): 31–49.

Harvey, David. 1989. *The condition of postmodernity*. Oxford: Oxford University Press.

1996. *Justice, nature and the geography of difference*. Oxford: Blackwell.

2018. *Marx, capital, and the madness of economic reason*. New York: Oxford University Press.

Headrick, Daniel R. 1981. *The tools of empire: Technology and European imperialism in the nineteenth century*. Oxford: Oxford University Press.

2009. *Technology: A world history*. Oxford: Oxford University Press.

2010. *Power over peoples: Technology, environments, and Western imperialism, 1400 to the present*. Princeton, NJ: Princeton University Press.

Heffron, Raphael J., and Darren McCauley. 2017. The concept of energy justice across the disciplines. *Energy Policy* 105: 658–667.

Heilbroner, Robert. [1953] 1999. *The worldly philosophers: The lives, times and ideas of the great economic thinkers*. London: Penguin.

Hickman, Larry A., and Elizabeth F. Porter, eds. 1993. *Technology and ecology: The proceedings of the VII International Conference of the Society for Philosophy and Technology*. Carbondale, IL: The Society for Philosophy and Technology.

Hine, Dougald, and Paul Kingsnorth. 2009. Uncivilisation: The Dark Mountain Manifesto. *The Dark Mountain Project*. http://dark-mountain.net/about/manifesto/

Holleman, Hannah. 2018. *Dust bowls of empire: Imperialism, environmental politics, and the injustice of "green" capitalism*. New Haven, CT: Yale University Press.

Hornborg, Alf. 1992. Machine fetishism, value, and the image of unlimited good: Toward a thermodynamics of imperialism. *Man (New Series)* 27: 1–18.

1994. Environmentalism, ethnicity and sacred places: Reflections on modernity, discourse and power. *Canadian Review of Sociology and Anthropology* 31: 245–267.

1998a. Towards an ecological theory of unequal exchange: Articulating world system theory and ecological economics. *Ecological Economics* 25(1): 127–136.

1998b. Ecosystems and world systems: Accumulation as an ecological process. *Journal of World-Systems Research* 4(2): 169–177.

2001. *The power of the machine: Global inequalities of economy, technology, and environment*. Walnut Creek, CA: AltaMira.

2002. Review of Tim Ingold, "The perception of the environment: Essays in livelihood, dwelling and skill." *Ethnos* 67(1): 121–122.

2006. Footprints in the cotton fields: The Industrial Revolution as time-space appropriation and environmental load displacement. *Ecological Economics* 59(1): 74–81.

2010. Toward a truly global environmental history: A review article. *Review* 33 (4): 295–323.

2013a. *Global ecology and unequal exchange: Fetishism in a zero-sum world.* Revised paperback version. London: Routledge.

2013b. The fossil interlude: Euro-American power and the return of the Physiocrats. In *Cultures of energy: Power, practices, technologies*, edited by Sarah Strauss, Stephanie Rupp, and Thomas Love, 41–59. Walnut Creek, CA: Left Coast Press.

2015. The political ecology of the Technocene: Uncovering ecologically unequal exchange in the world-system. In *The Anthropocene and the global environmental crisis: Rethinking modernity in a new epoch*, edited by Clive Hamilton, Christophe Bonneuil, and Francois Gemenne, 57–69. London: Routledge.

2016. *Global magic: Technologies of appropriation from ancient Rome to Wall Street.* Houndmills, UK: Palgrave Macmillan.

Hornborg, Alf, and Carole L. Crumley, eds. 2007. *The world system and the Earth system: Global socioenvironmental change and sustainability since the Neolithic.* Walnut Creek, CA: Left Coast Press.

Hornborg, Alf, and Joan Martinez-Alier, eds. 2016. Ecologically unequal exchange and ecological debt. Special section of the *Journal of Political Ecology* 23(1): 328–491.

Hornborg, Alf, John R. McNeill, and Joan Martinez-Alier, eds. 2007. *Rethinking environmental history: World-system history and global environmental change.* Lanham, MD: AltaMira.

Huber, Matthew T. 2008. Energizing historical materialism: Fossil fuels, space and the capitalist mode of production. *Geoforum* 40: 105–115.

2013. *Lifeblood: Oil, freedom, and the forces of capital.* Minneapolis: University of Minnesota Press.

2015. Theorizing energy geographies. *Geography Compass* 9(6): 327–338.

Huber, Matthew T., and James McCarthy. 2017. Beyond the subterranean energy regime? Fuel, land use and the production of space. *Transactions of the Institute of British Geographers* 42(4): 655–668.

Huesemann, Michael, and Joyce Huesemann. 2011. *Techno-fix: Why technology won't save us or the environment.* Gabriola Island, BC: New Society Publishers.

Hughes, Thomas P. 1987. The evolution of large technological systems. In *The social construction of technological systems: New directions in the sociology and history of technology*, edited by Wiebe E. Bijker, Thomas P. Hughes, and Trevor J. Pinch, 45–76. Cambridge, MA: MIT Press.

Ihde, Don. 1993. *Philosophy of technology: An introduction.* New York: Paragon House.

2010. Philosophy of technology (and/or technoscience?): 1996–2010. *Techné* 14(1): 26–35.

Ihde, Don, and Evan Selinger, eds. 2003. *Chasing technoscience: Matrix for materiality*. Bloomington: Indiana University Press.

Illich, Ivan. 1973. *Tools for conviviality*. London: Calder and Boyars.

1983. The social construction of energy. Opening talk at seminar on "The basic option within any future low-energy society," El Colegio de México, July.

Ingold, Tim. 1988. Tools, minds and machines: An excursion in the philosophy of technology. *Techniques et Culture* 12: 151–176.

1997. Eight themes in the anthropology of technology. *Social Analysis: The International Journal of Social and Cultural Practice* 41(1): 106–138.

2000. *The perception of the environment: Essays in livelihood, dwelling and skill*. London: Routledge.

2015. *The life of lines*. London: Routledge.

2018. From science to art and back again: The pendulum of an anthropologist. *Interdisciplinary Science Reviews* 43(3–4): 213–227.

Inikori, Joseph E. 1989. Slavery and the revolution in cotton textile production in England. *Social Science History* 13(4): 343–379.

2002. *Africans and the Industrial Revolution in England: A study of international trade and economic development*. Cambridge: Cambridge University Press.

Jackson, Tim. 2009. *Prosperity without growth: Economics for a finite planet*. London: Earthscan.

Jenkins, Kirsten, Darren McCauley, Raphael Heffron, Hannes Stephan, and Robert Rehner. 2016. Energy justice: A conceptual review. *Energy Research and Social Science* 11: 174–182.

Jorgenson, Andrew K., and Brett Clark, eds. 2009. Special issue: Ecologically unequal exchange in comparative perspective. *International Journal of Comparative Sociology* 50(3–4): 211–409.

Kallis, Giorgos. 2011. In defence of degrowth. *Ecological Economics* 70: 873–880.

2018. *Degrowth*. Newcastle upon Tyne, UK: Agenda Publishing.

Keen, Steve. 1993a. Use-value, exchange value, and the demise of Marx's labor theory of value. *Journal of the History of Economic Thought* 15(1): 107–121.

1993b. The misinterpretation of Marx's theory of value. *Journal of the History of Economic Thought* 15(2): 282–300.

[2001] 2011. *Debunking economics: The naked emperor dethroned?* London: Zed Books.

Kipnis, Andrew B. 2014. Agency between humanism and posthumanism: Latour and his opponents. *HAU: Journal of Ethnographic Theory* 5(2): 43–58.

Kirchhoff, Michael D. 2009. Material agency: A theoretical framework for ascribing agency to material culture. *Techné* 13(3): 206–220.

Klein, Naomi. 2014. *This changes everything: Capitalism vs. the climate*. London: Allen Lane.

Kohn, Eduardo. 2013. *How forests think: Toward an anthropology beyond the human*. Berkeley: University of California Press.

Kolbert, Elizabeth. 2014. *The sixth extinction: An unnatural history.* New York: Picador.

Koumoundouros, Tessa. 2018. To save ourselves it's time to rethink our economic system, warn scientists. *Science Alert*, September 10.

Kovel, Joel. 2002. *The enemy of nature: The end of capitalism or the end of the world?* London: Zed Books.

Lash, Scott, and John Urry. 1994. *Economies of signs and space.* London: Sage.

Latouche, Serge. 2009. *Farewell to growth.* Cambridge: Polity Press.

Latour, Bruno. 1987. *Science in action: How to follow scientists and engineers through society.* Cambridge, MA: Harvard University Press.

   1993a. *We have never been modern.* Cambridge, MA: Harvard University Press.

   1993b. Ethnography of a "high-tech" case: About Aramis. In *Technological choices: Transformation in material cultures since the Neolithic*, edited by Pierre Lemonnier, 372–398. London: Routledge.

   1996. *Aramis or the love of technology.* Cambridge, MA: Harvard University Press.

   2004a. *Politics of nature: How to bring the sciences into democracy.* Cambridge, MA: Harvard University Press.

   2004b. Why has critique run out of steam? From matters of fact to matters of concern. *Critical Inquiry* 30: 225–248.

   2005. *Reassembling the social: An introduction to Actor-Network-Theory.* Oxford: Oxford University Press.

   2010. *On the modern cult of the factish gods.* Durham, NC: Duke University Press.

   2014. Agency at the time of the Anthropocene. Holberg Prize Lecture. *New Literary History* 45: 1–18.

   2016. How not to be too mistaken about Trump? *Le Monde*, November 12–13. Translated by Clara Soudan and Jaeyoon Park.

   2017. *Facing Gaia: Eight lectures on the new climatic regime.* Cambridge: Polity.

   2018. *Down to earth: Politics in the new climatic regime.* Cambridge: Polity Press.

Latour, Bruno, and Steve Woolgar. 1979. *Laboratory life: The construction of scientific facts.* Beverly Hills, CA: Sage.

Law, John. 1987. Technology and heterogeneous engineering: The case of Portuguese expansion. In *The social construction of technological systems: New directions in the sociology and history of technology*, edited by Wiebe E. Bijker, Thomas P. Hughes, and Trevor J. Pinch, 105–127. Cambridge, MA: MIT Press.

   2015. What's wrong with a one-world world? *Distinktion: Journal of Social Theory* 16(1): 126–139.

Lawson, Clive. 2008. An ontology of technology: Artefacts, relations and functions. *Techné* 12(1): 48–64.

Layton, Edwin T., Jr. 1974. Technology as knowledge. *Technology and Culture* 15(1): 31–41.

Le Goff, Jacques. 2012. *Money in the Middle Ages.* Cambridge: Polity.

Lemonnier, Pierre, ed. 1993. *Technological choices: Transformation in material cultures since the Neolithic*. London: Routledge.

Lenzen, Manfred, Keiichiro Kanemoto, Daniel Moran, and Arne Geschke. 2012. Mapping the structure of the world economy. *Environmental Science and Technology* 46(15): 8374–8381.

Lenzen, Manfred, Daniel Moran, Keiichiro Kanemoto, and Arne Geschke. 2013. Building EORA: A global multi-regional input-output database at high country and sector resolution. *Economic Systems Research* 25(1): 20–49.

Lipietz, Alain. 2000. Political ecology and the future of Marxism. *Capitalism Nature Socialism* 11: 69–85.

Locher, Fabien, and Jean-Baptiste Fressoz. 2012. Modernity's frail climate: A climate history of environmental reflexivity. *Critical Inquiry* 38(3): 579–598.

Lomborg, Björn. 2001. *The skeptical environmentalist: Measuring the real state of the world*. Cambridge: Cambridge University Press.

Lonergan, Stephen C. 1988. Theory and measurement of unequal exchange: A comparison between a Marxist approach and an energy theory of value. *Ecological Modeling* 41: 127–145.

Lovelock, James. 2000. *Gaia: The practical science of planetary medicine*. Oxford: Oxford University Press.

Macfarlane, Alan. 1985. The root of all evil. In *The anthropology of evil*, edited by David Parkin, 57–76. Oxford: Blackwell.

Malm, Andreas. 2016. *Fossil capital: The rise of steam power and the roots of global warming*. London: Verso.

2018. *The progress of this storm: Nature and society in a warming world*. London: Verso.

Malm, Andreas, and Alf Hornborg. 2014. The geology of mankind? A critique of the Anthropocene narrative. *The Anthropocene Review* 1: 62–69.

Maris, Virginie. 2015. Back to the Holocene: A conceptual, and possibly practical, return to a nature not intended for humans. In *The Anthropocene and the global environmental crisis: Rethinking modernity in a new epoch*, edited by Clive Hamilton, Christophe Bonneuil, and Francois Gemenne, 123–133. London: Routledge.

Marks, Robert B. [2002] 2015. *The origins of the modern world: A global and environmental narrative from the fifteenth to the twenty-first century*. Lanham, MD: Rowman and Littlefield.

Martin, Keir. 2014. Afterword: Knot-work not networks, or anti-anti-antifetishism and the ANTipolitics machine. *HAU: Journal of Ethnographic Theory* 4 (3): 99–115.

Martinez-Alier, Joan. 1987. *Ecological economics: Energy, environment and society*. Oxford: Blackwell.

2002. *The environmentalism of the poor: A study of ecological conflicts and valuation*. Cheltenham, UK: Edward Elgar.

Martinez-Alier, Joan, and José M. Naredo. 1982. A Marxist precursor of energy economics: Podolinsky. *The Journal of Peasant Studies* 9(2): 207–224.

Marx, Karl. [1857] 1973. *Grundrisse: Foundations of the critique of political economy*. London: Penguin.

[1867] 1967. *Capital*, vol.1. New York: International Publishers.

[1867] 1976. *Capital*, vol.1. Harmondsworth, UK: Penguin Books.

Mauss, Marcel. [1925] 2016. *The gift: The form and reason for exchange in archaic societies*. Translated by Jane I. Guyer. Chicago: HAU Books.

McNeill, John R., and Peter Engelke. 2016. *The Great Acceleration: An environmental history of the Anthropocene since 1945*. Cambridge, MA: The Belknap Press of Harvard University Press.

Meadows, Donella H., Dennis L. Meadows, Jorgen Randers, and William W. Behrens III. 1972. *Limits to growth*. New York: New American Library.

Mellor, Mary. 2005. The politics of money and credit as a route to ecological sustainability and economic democracy. *Capitalism Nature Socialism* 16(2): 45–60.

2009. The financial crisis. *Capitalism Nature Socialism* 20(1): 34–36.

Millennium Ecosystem Assessment. 2005. *Living beyond our means: Natural assets and human well-being*. Washington, DC: World Resources Institute.

Miller, Clark. 2014. The ethics of energy transitions. Paper presented at IEEE International Symposium on Ethics in Science, Technology and Engineering. *Institute of Electrical and Electronics Engineers*.

Miller, Daniel. 1987. *Material culture and mass consumption*. Oxford: Blackwell Publishers.

Miller, Daniel, ed. 2005. *Materiality*. Durham, NC: Duke University Press.

Mirowski, Philip. 1988. Energy and energetics in economic theory: A review essay. *Journal of Economic Issues* 22(3): 811–830.

1989. *More heat than light: Economics as social physics, physics as nature's economics*. New York: Cambridge University Press.

2013. *Never let a serious crisis go to waste: How neoliberalism survived the financial meltdown*. London: Verso.

Mitcham, Carl. 1994. *Thinking through technology: The path between engineering and philosophy*. Chicago: University of Chicago Press.

Mitchell, Timothy. 2011. *Carbon democracy: Political power in the age of oil*. London: Verso.

Moore, Jason W. 2000. Marx and the historical ecology of capital accumulation on a world scale: A comment on Alf Hornborg's "Ecosystems and World Systems: Accumulation as an Ecological Process." *Journal of World-Systems Research* VI(1): 133–138.

2007. Silver, ecology, and the origins of the modern world, 1450–1640. In *Rethinking environmental history: World-system history and global environmental change*, edited by Alf Hornborg, John R. McNeill, and Joan Martinez-Alier, 123–142. Lanham, MD: AltaMira.

2015. *Capitalism in the web of life: Ecology and the accumulation of capital*. London: Verso.

2016. The rise of cheap nature. In *Anthropocene or Capitalocene? Nature, history, and the crisis of capitalism*, edited by Jason W. Moore, 78–115. Oakland, CA: PM Press.

Moore, Jason W., ed. 2016. *Anthropocene or Capitalocene? Nature, history, and the crisis of capitalism*. Oakland, CA: PM Press.

Morris, Rosalind C., and Daniel H. Leonard. 2017. *The returns of fetishism: Charles de Brosses and the afterlives of an idea.* Chicago: University of Chicago Press.

Moseley, Fred. 2016. *Money and totality: A macro-monetary interpretation of Marx's logic in Capital and the end of the "Transformation Problem."* Leiden, The Netherlands: Brill.

Moseley, Fred, ed. 2005. *Marx's theory of money: Modern appraisals.* Houndmills, UK: Palgrave Macmillan.

Munn, Nancy. 1986. *The fame of Gawa.* Cambridge: Cambridge University Press.

Nelson, Anitra. 1999. *Marx's concept of money: The god of commodities.* London: Routledge.

———. 2001a. The poverty of money: Marxian insights for ecological economists. *Ecological Economics* 36: 499–511.

———. 2001b. Marx's theory of the money commodity. *History of Economics Review* 33: 44–63.

Nelson, Anitra, and Frans Timmerman, eds. 2011. *Life Without Money: Building Fair and Sustainable Economies.* London: Pluto Press.

Nikiforuk, Andrew. 2012. *The energy of slaves: Oil and the new servitude.* Vancouver, BC: Greystone Books.

Nixon, Robert. 2011. *Slow violence and the environmentalism of the poor.* Cambridge, MA: Harvard University Press.

Noble, David F. 1999. *The religion of technology: The divinity of Man and the spirit of invention.* London: Penguin Books.

Norberg, Johan. 2016. *Progress: Ten reasons to look forward to the future.* London: Oneworld Publications.

North, Peter. 2007. *Money and liberation: The micropolitics of alternative currency movements.* Minneapolis: University of Minnesota Press.

Nye, David E. 2006. *Technology matters: Questions to live with.* Cambridge, MA: MIT Press.

O'Connor, James. 1988. Capitalism, nature, socialism: A theoretical introduction. *Capitalism Nature Socialism* 1(1): 11–38.

———. 1998. *Natural causes: Essays in ecological Marxism.* New York: Guilford Press.

O'Connor, Martin, ed. 1994. *Is capitalism sustainable? Political economy and the politics of ecology.* New York: Guilford Press.

Odum, Howard T. 1971. *Environment, power, and society.* New York: Wiley-Interscience.

———. 1988. Self-organization, transformity, and information. *Science* 242: 1132–1139.

———. 1996. *Environmental accounting: Emergy and environmental decision making.* New York: John Wiley and Sons.

Odum, Howard T., and Jan E. Arding. 1991. *Emergy analysis of shrimp mariculture in Ecuador.* Working Paper, Coastal Resources Center, University of Rhode Island, Narragansett.

Oreskes, Naomi, and Erik M. Conway. 2008. Challenging knowledge: How climate science became a victim of the Cold War. In *Agnotology: The making and unmaking of ignorance,* edited by Robert N. Proctor and Londa Schiebinger, 55–89. Stanford, CA: Stanford University Press.

Parenti, Christian. 2016. Environment-making in the Capitalocene: Political ecology of the State. In *Anthropocene or Capitalocene? Nature, history, and the crisis of capitalism*, edited by Jason W. Moore, 166–184. Oakland, CA: PM Press.

Peet, Richard, Paul Robbins, and Michael J. Watts, eds. 2011. *Global political ecology*. London: Routledge.

Pérez Rincón, Mario. 2006. Colombian international trade from a physical perspective: Towards an ecological "Prebisch Thesis." *Ecological Economics* 59: 519–529.

Perreault, Tom, Gavin Bridge, and James McCarthy, eds. 2015. *The Routledge handbook of political ecology*. London: Routledge.

Pfaffenberger, Bryan. 1988a. Fetishised objects and humanized nature: Towards an anthropology of technology. *Man (New Series)* 23(2): 236–252.

1988b. The Hindu *Akama* temple as a machine, or, the Western machine as a temple. Paper presented at the Department of Cultural Anthropology, Uppsala University.

1992. Social anthropology of technology. *Annual Reviews in Anthropology* 21: 491–516.

Piketty, Thomas. 2014. *Capital in the twenty-first century*. Cambridge, MA: Belknap Press.

Pimentel, David, Tad Patzek, and Gerald Cecil. 2007. Ethanol production: Energy, economic, and environmental losses. *Reviews of Environmental Contamination and Toxicology* 189: 25–41.

Pinch, Trevor J., and Wiebe E. Bijker. 1987. The social construction of facts and artifacts: Or how the sociology of science and the sociology of technology might benefit each other. In *The social construction of technological systems: New directions in the sociology and history of technology*, edited by Wiebe E. Bijker, Thomas P. Hughes, and Trevor J. Pinch, 11–44. Cambridge, MA: MIT Press.

Pinker, Steven. 2018. *Enlightenment now: The case for reason, science, humanism and progress*. London: Allen Lane.

Podolinsky, Sergei. [1883] 2008. Human labor and unity of force. *Historical Materialism* 16: 163–183.

Polanyi, Karl. [1944] 1957. *The great transformation: The political and economic origins of our time*. Boston: Beacon.

Pomeranz, Kenneth. 2000. *The great divergence: China, Europe, and the making of the modern world economy*. Princeton, NJ: Princeton University Press.

Post, Robert C. 2010. Back at the start: History and technology and culture. *Technology and Culture* 51(4): 961–994.

Pouillon, Jean. [1982] 2016. Remarks on the verb "to believe." *HAU: Journal of Ethnographic Theory* 6(3): 485–492.

Prieto, Pedro A., and Charles A. S. Hall. 2013. *Spain's photovoltaic revolution: The energy return on investment*. New York: Springer.

Proctor, Robert N., and Londa Schiebinger, eds. 2008. *Agnotology: The making and unmaking of ignorance*. Stanford, CA: Stanford University Press.

Rabinbach, Anson. 1990. *The human motor: Energy, fatigue and the origins of modernity*. New York: Basic Books.

Rappaport, Roy A. 1994. Humanity's evolution and anthropology's future. In *Assessing cultural anthropology*, edited by Robert Borofsky, 152–166. New York: McGraw-Hill.

Raworth, Kate. 2017. *Doughnut economics: Seven ways to think like a 21st century economist*. White River Junction, VT: Chelsea Green Publishing.

Ripple, William J. et al. 2017. World scientists' warning to humanity: A second notice. *BioScience* 67(12): 1026–1028.

Robbins, Paul. 2012. *Political ecology: A critical introduction*. Malden, MA: Wiley-Blackwell.

Roberts, J. Timmons, and Bradley C. Parks. 2007. *A climate of injustice: Global inequality, North-South politics, and climate policy*. Cambridge, MA: MIT Press. 2008. Fueling injustice: Globalization, ecologically unequal exchange and climate change. *Globalizations* 4(2): 193–210.

Rockström, Johan, Will Steffen, Kevin Noone, and Åsa Persson. 2009. A safe operating space for humanity. *Nature* 461: 472–475.

Rosling, Hans, Anna Rosling Rönnlund, and Ola Rosling. 2018. *Factfulness: Ten reasons we're wrong about the world – and why things are better than you think*. New York: Flatiron Books.

Rubel, Maximilien, and John, Crump. 1987. *Non-Market Socialism in the Nineteenth and Twentieth Centuries*. New York: Palgrave Macmillan.

Rudy, Alan. 2001. *Marx's Ecology* and rift analysis. *Capitalism Nature Socialism* 12(2): 56–63.

Ruel, Malcolm. [1982] 2002. Christians as believers. In *A reader in the anthropology of religion*, edited by Michael Lambek, 99–113. Malden, MA: Blackwell.

Ruggiero, Vincenzo. 2013. *The crimes of the economy: A criminological analysis of economic thought*. London: Routledge.

Sahlins, Marshall D. 1976. *Culture and practical reason*. Chicago: University of Chicago Press.

Sale, Kirkpatrick. 1995. *Rebels against the future: The Luddites and their war on the Industrial Revolution*. New York: Basic Books.

Salleh, Ariel. 1994. Nature, woman, labor, capital: Living the deepest contradiction. In *Is capitalism sustainable? Political economy and the politics of ecology*, edited by Martin O'Connor, 106–124. New York: Guilford.

Sanders, Bernie. 2015. *The speech: On corporate greed and the decline of our middle class*. New York: Nation Books.

Santos-Granero, Fernando, ed. 2009. *The occult life of things: Native Amazonian theories of materiality and personhood*. Tucson: University of Arizona Press.

Schandl, Heinz, M. Fischer-Kowalski, J. West, S. Giljum, M. Dittrich, N. Eisenmenger, A. Geschke, M. Lieber, H.P. Wieland, A. Schaffartzik, F. Krausmann, S. Gierlinger, K. Hosking, M. Lenzen, H. Tanikawa, A. Miatto, and T. Fishman. 2016. *Global material flows and resource productivity: Assessment Report for the UNEP International Resource Panel*. Paris: United Nations Environment Programme.

Scharff, Robert C., and Val Dusek, eds. 2014. *Philosophy of technology: The technological condition – an anthology*. Malden, MA: Wiley Blackwell.

Scharlemann, Jörn P. W., and William F. Laurance, 2008. How green are biofuels? *Science* 319: 43–44.

Schumacher, Ernst F. 1973. *Small is beautiful: A study of economics as if people mattered.* New York: Harper and Row.

——— 1979. *Good work.* London: Jonathan Cape.

Schwartzman, David. 1996. Solar communism. *Science and Society* 60(3): 307–331.

Seaquist, Jonathan W., Emma L. Johansson, and Kimberly A. Nicholas. 2014. Architecture of the global land acquisition system: Applying the tools of network science to identify key vulnerabilities. *Environmental Research Letters* 9: 114006.

Sieber, Nikolai I. [1871] 2001. Marx's theory of value and money: Marx's *Capital* and capitalism. *Research in Political Economy* 19: 17–45.

Sieferle, Rolf Peter. 2001. *The subterranean forest: Energy systems and the Industrial Revolution.* Cambridge: White Horse Press.

Sillar, Bill. 2009. The social agency of things? Animism and materiality in the Andes. *Cambridge Archaeological Journal* 19(3): 367–377.

Simas, Moana, Richard Wood, and Edgar Hertwich. 2015. Labor embodied in trade: The role of labor and energy productivity and implications for greenhouse gas emissions. *Journal of Industrial Ecology* 19(3): 343–356.

Simmel, Georg. [1907] 1990. *The philosophy of money.* London: Routledge.

Smil, Vaclav. 2015. *Power density: A key to understanding energy sources and uses.* Cambridge, MA: MIT Press.

"Smith," A. [alias Goodman, G. J. W.] 1967. *The money game.* New York: Vintage.

Smith, Merritt R., and Leo Marx, eds. 1994. *Does technology drive history? The dilemma of technological determinism.* Cambridge, MA: MIT Press.

Smith, Neil. 1984. *Uneven development: Nature, capital, and the production of space.* New York: Blackwell.

Soper, Kate. 1991. Greening Prometheus. In *Socialism and the limits of liberalism,* edited by Peter Osborne, 271–293. London: Verso.

——— 1995. *What is Nature? Culture, politics and the non-human.* Oxford: Blackwell.

——— 2012. The humanism in posthumanism. *Comparative Critical Studies* 9(3): 365–378.

Sovacool, Benjamin K., and Michael H. Dworkin. 2015. Energy justice: Conceptual insights and practical applications. *Applied Energy* 142: 435–444.

Spash, Clive L., ed. 2017. *Routledge handbook of ecological economics: Nature and society.* London: Routledge.

Spier, Robert F. G. 1970. *From the hand of Man: Primitive and preindustrial technologies.* Boston: Houghton Mifflin Company.

Spyer, Patricia, ed. 1998. *Border fetishisms: Material objects in unstable spaces.* London: Routledge.

Staudenmaier, John M. 1984. What SHOT hath wrought and what SHOT hath not: Reflections on twenty-five years of the history of technology. *Technology and Culture* 25(4): 707–730.

Steffen, Will, Wendy Broadgate, Lisa Deutsch, Owen Gaffney, and Cornelia Ludwig. 2015. The trajectory of the Anthropocene: The Great Acceleration. *The Anthropocene Review* 2(1): 81–98.

Steffen, Will et al. 2004. *Global change and the Earth system: A planet under pressure.* The IGBP Book Series. Berlin: Springer.

Stern, Paul C., Benjamin K. Sovacool, and Thomas Dietz. 2016. Towards a science of climate and energy choices. *Nature Climate Change* 6: 547–555.

Stover, Carl F., ed. 1962. *The Encyclopedia Britannica Conference on the Technological Order*. Special issue of *Technology and Culture*.

Strauss, Sarah, Stephanie Rupp, and Thomas Love, eds. *Cultures of energy: Power, practices, technologies*. Walnut Creek, CA: Left Coast Press.

Summers, Lawrence. 1991. Memorandum. www.whirledbank.org/ourwords/summers.html.

Strum, Shirley S., and Bruno Latour. 1987. Redefining the social link: From baboons to humans. *Social Science Information* 26(4): 783–802.

Tainter, Joseph A. 1988. *The collapse of complex societies*. Cambridge: Cambridge University Press.

Tainter, Joseph A., and Tad W. Patzek. 2012. *Drilling down: The gulf oil debacle and our energy dilemma*. New York: Springer.

Tambiah, Stanley J. 1990. *Magic, science, religion, and the scope of rationality*. Cambridge: Cambridge University Press.

Tawney, Richard H. 1972. *Religion and the rise of capitalism*. Harmondsworth, UK: Penguin.

Tsing, Anna L. 2015. *The mushroom at the end of the world: On the possibility of life in capitalist ruins*. Princeton, NJ: Princeton University Press.

Tyfield, David, and John Urry, eds. 2014. Special Issue on Energy and Society. *Theory, Culture and Society* 31(3).

Union of Concerned Scientists. 1992. *World scientists' warning to humanity*. Cambridge, MA: Union of Concerned Scientists.

Vettese, Troy. 2018. To freeze the Thames: Natural geo-engineering and biodiversity. *New Left Review* 111: 63–86.

Victor, Peter A. 2008. *Managing without growth: Slower by design, not disaster*. Cheltenham, UK: Edward Elgar.

Victor, Peter A., and Tim Jackson. 2015. The trouble with growth. In *State of the World 2015: Confronting hidden threats to sustainability*, 37–49. The Worldwatch Institute. Washington, DC: Island Press.

Viveiros de Castro, Eduardo. 1998. Cosmological deixis and Amerindian perspectivism. *The Journal of the Royal Anthropological Institute* 4(3): 469–488.

von Uexküll, Jakob. [1940] 1982. The theory of meaning. *Semiotica* 42: 25–82.

2010. *A foray into the worlds of animals and humans, with a theory of meaning*. Minneapolis: University of Minnesota Press.

Wallace-Wells, David. 2017. The uninhabitable Earth. *New York*, July 10. http://nymag.com/daily/intelligencer/2017/07/climate-change-earth-too-hot-for-humans.html.

Wallerstein, Immanuel. 1974–1989. *The modern world-system 1–3*. San Diego, CA: Academic Press.

2004. *World-systems analysis: An introduction*. Durham, NC: Duke University Press.

Warlenius, Rikard. 2011. *Iron for tea: The trade of the Swedish East India Company as a cross-continental case study of ecologically unequal exchange in the eighteenth century*. Master's Thesis, Department of Economic History, Stockholm University.

Watts, Michael J. 2015. Now and then: The origins of political ecology and the rebirth of adaptation as a form of thought. In *The Routledge handbook of political ecology*, edited by Tom Perreault, Gavin Bridge, and James McCarthy, 19–50. London: Routledge.

Weatherford, Jack. 1997. *The history of money: From sandstone to cyberspace.* New York: Three Rivers.

White, Leslie A. 1940. The symbol: The origin and basis of human behavior. *Philosophy of Science* 7(4): 451–463.

Wiener, Margaret J. 2013. Magic, (colonial) science and science studies. *Social Anthropology* 21 (4): 492–509.

Wilkinson, Richard G. 1973. *Poverty and progress: An ecological model of economic development.* London: Methuen.

Wilson, Edward O. 1984. *Biophilia.* Cambridge, MA: Harvard University Press.

2016. *Half-earth: Our planet's fight for life.* London: Liveright Publishing.

Williamsson, Phil. 2016. Scrutinize CO2 removal methods. *Nature* 530: 153–155.

Winner, Langdon. 1977. *Autonomous technology: Technics-out-of-control as a theme in political thought.* Cambridge, MA: MIT Press.

1980. Do artifacts have politics? *Daedalus* 109(1): 121–136.

1986. *The whale and the reactor: A search for limits in an age of high technology.* Chicago: University of Chicago Press.

1993. Upon opening the black box and finding it empty: Social constructivism and the philosophy of technology. *Science, Technology, and Human Values* 18(3): 362–378.

Wolf, Eric R. 1972. Ownership and political ecology. *Anthropological Quarterly* 45(3): 201–205.

1982. *Europe and the people without history.* Berkeley: University of California Press.

World Bank. 2012. *Turn down the heat: Why a 4 degree warmer world must be avoided.* Washington, DC: The World Bank.

World Economic Forum. 2014. *Global risks 2014.* Geneva: World Economic Forum.

World Wide Fund for Nature (WWF). 2016. *Living planet report 2016.* Gland, Switzerland: WWF.

Wrigley, E. Anthony. 1988. *Continuity, chance and change: The character of the Industrial Revolution in England.* Cambridge: Cambridge University Press.

Yu, Yang, Kuishuang Feng, and Klaus Hubacek. 2013. Tele-connecting local consumption to global land use. *Global Environmental Change* 23(5): 1178–1186.

Zelizer, Viviana A. 1998. The proliferation of social currencies. In *The laws of the markets*, edited by Michel Callon, 58–68. Oxford: Blackwell Publishers.

2017. *The social meaning of money.* Princeton, NJ: Princeton University Press.

Names Index

Albornoz, Cristóbal de, 213
Alexander, Samuel, 236
Altvater, Elmar, 172, 201–204
Alvaredo, Facundo, 41, 87, 92, 151, 251
Amin, Samir, 143n9
Andersen, Otto, 125
Angus, Ian, 42, 43, 45–47
Aquinas, Thomas (saint), 142, 172
Arding, Jan E., 58, 142
Arendt, Hannah, 198, 199
Aristotle, 27–28, 68, 80, 160–161, 164n7, 172
Asafu-Adjaye, John, 82, 147

Bailey, Sinead, 60
Barca, Stefania, 172
Barth, Fredrik, 56n2
Basso, Keith, 212
Bateson, Gregory, 4
Baudrillard, Jean, 160, 161, 170, 220
Bauman, Zygmunt, 55, 65, 75n11, 89
Beck, Ulrich, 125
Beckert, Sven, 94, 116
Benedict, Ruth, 103
Bennett, Jane, 11, 181n4
Benton, Ted, 142, 163, 170n9
Berg, Maxine, 31, 171
Besley, Tim, 83
Bessire, Lucas, 187
Bhaskar, Roy, 181n7
Bijker, Wiebe E., 31, 94, 97n5, 99, 100n11, 103n13, 104n15, 106, 124
Bildt, Carl, 89

Bjerg, Ole, 166–167
Blaikie, Piers, 144
Blanc, Jérôme, 236
Blaser, Mario, 178n3, 186, 208
Blaut, James M. (Jim), 31, 95n1
Bloch, Maurice, 77
Bohannan, Paul, 1n1, 233–234
Bond, David, 187
Bonneuil, Christophe, 38, 42, 45, 47, 51, 204, 206
Boulding, Kenneth, 82, 104
Bourdieu, Pierre, 24
Braudel, Fernand, 74n10
Brennan, Teresa, 164
Bridge, Gavin, 60, 125
Brookfield, Harold, 144
Broome, John, 64, 64n7
Bryant, Raymond L., 60
Buber, Martin, 182
Buck, Holly Jean, 137, 183n9
Bunker, Stephen G., 7, 57–59, 63n5, 153, 167, 170
Burkett, Paul, 124, 127, 143, 145, 152n1, 157, 159, 160, 163–165, 170, 173–174, 204
Büscher, Bram, 183n9

Callon, Michel, 97n5, 106, 188n17
Carson, Rachel, 3
Castree, Noel, 165
Cecil, Gerald, 119
Chakrabarty, Dipesh, 190–191
Chase-Dunn, Christopher, 180

Subject Index

Printed in the United States
by Baker & Taylor Publisher Services